Contents

1	The metric system	5
2	Quantities, units and abbreviations in common use	6
3	Addition and subtraction	10
4	Multiplication and division	14
5	Decimals and fractions	19
6	Ratios	21
7	Percentages	23
8	Averages	29
9	The use of symbols	31
10	Formulae	34
11	Pythagoras' theorem and square root tables	40
12	Moments	50
13	Levers	56
14	Water pressure	65
15	Box's formula	71
16	Chezy formula	74
17	Logarithms	78
18	Temperature scales and conversion tables	88
19	Conversion tables	90
20	Answers	95

Mathematics for Plumbers

C. Barnes T.Eng. (CEI), FIOP

Plumbing and Heating Manager, formerly Course Tutor in Plumbing, North East Wales Institute of Higher Education, Kelsterton College

Hutchinson of L

Other books of interest from Hutchinson

Plumbing 1 and 2 *A. L. Townsend*
Metric Calculations for Building Craft Students 1 and 2 *F. L. Tabberer*

Hutchinson & Co (Publishers) Ltd
3 Fitzroy Square, London W1P 6JD

London Melbourne Sydney Auckland
Wellington Johannesburg and agencies
throughout the world

First published 1979

Set in IBM Press Roman by Preface Ltd, Salisbury, Wilts
Printed in Great Britain by The Anchor Press Ltd
and bound by Wm Brendon & Son Ltd,
both of Tiptree, Essex

ISBN 0 09 129231 X

1 The metric system

Some years ago the government made the decision to change the form of measurement in this country from the traditional, imperial system, to the metric system employed by most of the rest of the world. To grasp the fundamentals of this system it is important to erase the old system from our minds.

This is, of course, more easily said than done since we have been raised in a society conditioned to think in the imperial system – which has served us well but which must now go. We have become familiar with decimal coinage and temperatures, and weights and measures must now follow.

Because we as a nation reject change, we create difficulties for ourselves and we still tend to think in imperial measures – the only way to overcome this is to familiarize ourselves fully with the SI metric system.

This is difficult for an apprentice plumber who is taught to think metric at a technical institution but, immediately he goes to work, is confronted by the old system being used by people who either have not had sufficient opportunity to change or who reject SI units out of hand.

It is apparent, however, that no matter what our feelings are we must make the change.

2 Quantities, units and abbreviations in common use

We should now look at the names of the SI units used for measuring length, area, volume and capacity.

Table 1 gives:

Column 1 Quantities of length, area, volume capacity and so on.
Column 2 SI metric units in which the above are expressed.
Column 3 The abbreviations in which the units are usually written.

Note that litres are better expressed in full rather than abbreviated to l as this is easily confused with the digit one (1).

Table 1

Quantity	*SI Unit*	*Unit Abbreviation*
Linear	millimetre	mm
Linear	centimetre	cm
Linear	metre	m
Linear	kilometre	km
Area	millimetre squared	mm^2
Area	centimetre squared	cm^2
Area	metre squared	m^2
Area	kilometre squared	km^2
Volume	millimetre cubed	mm^3
Volume	centimetre cubed	cm^3
Volume	metre cubed	m^3
Volume	kilometre cubed	km^3
Capacity	litre	l (better written in full to avoid confusion with 1)
Time	second	s
Rate of flow	litre per second	l/s

Quantity	*SI Unit*	*Unit Abbreviation*
Volume rate of flow	cubic metre per second	m^3/s
Pressure	newton per square metre	N/m^2
Pressure	newton per square millimetre	N/mm^2
Pressure	kilonewton per square metre	kN/m^2
Pressure	kilonewton per square millimetre	kN/mm^2
Pressure	meganewton per square metre	MN/m^2
Pressure	bar (100 kN/m^2)	bar
Pressure	millibar (100 N/m^2)	mb
Pressure	pascal (1 N/m^2)	p
Force	newton (derived from kgm/2)	N
Energy/work	joule (derived from kgm/2 or Nm)	J
Quantity of heat	joule	J
Power	watt (derived from Nm/s or J/s)	W
Electrical current	amperes	A
Electrical potential	volt (derived from W/A)	V
Electrical resistance	ohm (derived from V/A)	Ω
Luminosity	candela	cd
Temperature	degree Celsius (normal temperature use)	C
Temperature	degree Kelvin	K

Now that we have an idea of what to call the various quantities we can now examine the units a little more closely. The units are made up of multiples and sub-multiples and these fall under the headings of base unit, prefix, symbol and multiplication factor:

Base unit is the quantity which is taken as the base for a metric measurement. In weight the base unit is the gram, in length the metre and so on.

Prefix is the name given to the increase or decrease from the base line or unit, e.g. kilo = 1000 times (or 10^3) the base unit.

Prefix symbol is the symbol or abbreviation given to the prefix, in this case k for kilo.

Multiplication factor is the value by which the base unit has been multiplied. For example, 10^3 means that the unit is formed by taking the base unit and multiplying it by 10 x 10 x 10 = base unit x 1000.

Conversely 10^{-3} is termed a sub-multiple and means that the unit is formed by dividing the base unit into 10 x 10 x 10 = 1000 units per base unit.

If 10^{-6} is used then the unit has been formed by dividing the base unit into 10 x 10 x 10 x 10 x 10 x 10 = 1 000 000 units per base unit.

Table 2 of multiples and sub-multiples gives the prefix, prefix symbol and multiplication factor of these as they vary from base unit.

Table 2

Prefix	*Prefix symbol*	*Multiplication factor*
tera-	T	10^{12}
giga-	G	10^9
mega-	M	10^6
kilo-	K	10^3
hecto-	h	10^2
	Base unit	
deci-	d	10^{-1}
centi-	c	10^{-2}
milli-	m	10^{-3}
micro-	u	10^{-6}
nano-	n	10^{-9}
pico-	p	10^{-12}

The base unit can be metres, grams, newtons, etc. The most common multiples and sub-multiples used are the mega, kilo, centi and milli, thus,

If the basic unit is metres (m), then
1 mega-metre (Mm) = 1 000 000 metres
1 mega-metre (Mm) = 1000 kilo-metres
1 kilo-metre (km) = 1000 metres
1 metre (base unit) (m) = 1000 milli-metres (mm)
1 metre (m) = 100 centi-metres (cm)
1 metre (m) = 10 deci-metres (dm)

It can be seen that the multiples and sub-multiples are in multiples of 10 decimal places and to avoid cumbersome numbers as 1 000 000 (mm) it is much simpler to use the multiplication factor, in this case:

1 000 000 metres = 1 x 10^6 metres

It will also be seen that any numbers that contain more than three digits are expressed by leaving a space and not using a comma:

10 metres
100 metres
1000 metres
10 000 metres
100 000 metres
1 000 000 metres

(As will be noted, 1000 is an exception to this rule as thousands are printed with neither comma *nor* space.)

Because the complete change over to the metric system will take many years, conversion of imperial measurements to metric is inevitable, the tables on pages 90–4 will be useful.

3 Addition and subtraction

Before any calculations may be attempted, a sound knowledge of the basic operations (addition, subtraction, multiplication and division) is necessary. Table 3 will help with some practice in addition and subtraction.

Table 3

0	1	2	3	4	5	6	7	8	9
1	2	3	4	5	6	7	8	9	10
2	3	4	5	6	7	8	9	10	11
3	4	5	6	7	8	9	10	11	12
4	5	6	7	8	9	10	11	12	13
5	6	7	8	9	10	11	12	13	14
6	7	8	9	10	11	12	13	14	15
7	8	9	10	11	12	13	14	15	16
8	9	10	11	12	13	14	15	16	17
9	10	11	12	13	14	15	16	17	18

Using the table

Select any digit down the left hand side of the table and then any digit along the top row. Where the two lines meet will be the answer.

Worked examples

1. Add $7 + 6$

Select the number 7 on the left hand column and 6 at the top. Run your finger along to the place where the two columns meet and you will find the answer is 13.

Answer $7 + 6 = 13$

To use the table for subtracting, reverse the operation.

2. Subtract 7 from 13

Find the number 7 in the top line and run your finger down the column to number 13. Now run your finger across to the left hand column where the answer will be 6.

Answer $13 - 7 = 6$

Worked examples

Addition

1. Add $6 + 7 + 3 + 4$

(look for the digits that add up to 10) then we have
$6 + 4$ and $7 + 3 = 10 + 10$

Answer 20

2. Add $5 + 3 + 6 + 5 + 7 + 4$
$= 5 + 5$ and $6 + 4$ and $7 + 3 = 10 + 10 + 10$

Answer 30

3. Add $1 + 5 + 7 + 4 + 5 + 2 + 3 + 6 + 8 + 9$
$= 1 + 9$ and $5 + 5$ and $7 + 3$ and $6 + 4$ and $2 + 8$
$= 10 + 10 + 10 + 10 + 10$

Answer 50

Not all figures are as conveniently grouped as the examples given but with a little practice you will be able to calculate additions like these more quickly.

4. $\begin{array}{r} 7 \\ +9 \\ \hline 16 \end{array}$

5. $\begin{array}{r} 43 \\ +17 \\ \hline 60 \end{array}$

6. $\begin{array}{r} 765 \\ +235 \\ \hline 1000 \end{array}$

7. $\begin{array}{r} 76453 \\ +23547 \\ \hline 100000 \end{array}$

8. $\begin{array}{r} 7 \\ 5 \\ 6 \\ 3 \\ 4 \\ +5 \\ \hline 30 \end{array}$

9. $\begin{array}{r} 9 \\ 7 \\ 5 \\ 1 \\ 4 \\ 3 \\ 6 \\ +5 \\ \hline 40 \end{array}$

Examples 1 to 9 deal with whole numbers only, so we will look at some examples that use parts of whole numbers or *decimals*. A *decimal point* separates the parts of the number from the whole. It is important that the point is placed in the correct position or the value of the answer will alter and be wrong. In the following calculations you can see the effect of placing the point in the wrong place when adding.

10.	11.	12.
6.7	6.7	6.7
66.8	668	66.8
12678	7.35	73.5

Two of these answers are wrong for the following reasons: in example 10, we ignored the decimal point altogether and the answer therefore became a whole number, 12678. In 11, we put the decimal point in the wrong place and got the answer 7.35. However, by placing the points one over the other in 12 the point was in the right place in the answer which was 73.5.

13. Add 1.1, 3.2, 6.5, 23.4, 171.1, 6 and 0.112

14. Add 2, 32, 67, 5, 12, 1.67, 15, 33, 35

You will immediately see that it will not be easy to calculate these figures in your head because of the decimal points that are included, so the sums should be set out as they were in Example 12.

13.	14.
1.1	2
3.2	32
6.5	67
23.4	5
171.1	12
6.0	1.67
0.112	15
211.412	33
	35
	202.67

Subtraction

1.	2.	3.	4.	5.	6.
10	20	47	20	45	50
−7	−17	−17	−14	−15	−25
3	3	30	6	30	25

The above simple examples should present no problems and can be checked by covering the top figure and adding together the bottom one and the answer.

For example

$$\begin{array}{r} 10 \\ -\underline{7} \\ 3 \end{array} \text{ cover the 10 and add } \begin{array}{r} 7 \\ \underline{+3} \\ 10 \end{array}$$

In the examples used so far whole numbers have, again, been used. Now we will use numbers with decimals.

7. $\begin{array}{r} 10.16 \\ -\underline{6.16} \\ 4.00 \end{array}$ 8. $\begin{array}{r} 10.16 \\ -\underline{6.10} \\ 4.06 \end{array}$ 9. $\begin{array}{r} 10.16 \\ -\underline{6.06} \\ 4.10 \end{array}$

You will notice that the decimal points are set one above the other as they were for addition. This is to ensure that the point will be in the correct place in the answer. The subtraction is otherwise unaltered.

10. $\begin{array}{r} 56\,789.432 \\ -\underline{12\,345.234} \\ 44\,444.198 \end{array}$

Revision questions

Addition

1. 7 + 3 2. 13 + 7 3. 5 + 4 + 2 4. 5 + 8 + 5
5. 17 + 5 + 3 + 5 6. 8.5 + 2 7. 7.5 + 0.5 + 2
8. 5 + 7 + 6 + 5 + 3 + 4 9. 5.4 + 3.6 + 2.7 + 1.3
10. 1.25 + 6.75 + 4.5

Subtraction

1. 10 – 7 2. 17 – 10 3. 45 – 15 4. 33 – 13
5. 21 – 7 6. 10.5 – 5 7. 19.50 – 9.25 8. 97 – 37
9. 2.6 – 0.6 10. 13.04 – 3.0

4 Multiplication and division

The multiplication and division table uses a similar system to the addition and subtraction table, but this time we find the answer to multiplying two digits together.

Table 4.

Multiplication and division (times) table

1	2	3	4	5	6	7	8	9	10	11	12
2	4	6	8	10	12	14	16	18	20	22	24
3	6	9	12	15	18	21	24	27	30	33	36
4	8	12	16	20	24	28	32	36	40	44	48
5	10	15	20	25	30	35	40	45	50	55	60
6	12	18	24	30	36	42	48	54	60	66	72
7	14	21	28	35	42	49	56	63	70	77	84
8	16	24	32	40	48	56	64	72	80	88	96
9	18	27	36	45	54	63	72	81	90	99	108
10	20	30	40	50	60	70	80	90	100	110	120
11	22	33	44	55	66	77	88	99	110	121	132
12	24	36	48	60	72	84	96	108	120	132	144

Example Multiply 7 by 6.
Solution Find the digit 7 on the top line and also the digit 6 in the left-hand column. Where the lines of the two digits meet we have the answer, 42.
Answer $7 \times 6 = 42$

To use the same table for division we go through the same operations in reverse.

Example Divide 42 by 7.
Solution Find the digit 42 under the column headed by the digit 7 on the top row. If we move from the digit 42 horizontally to the left we find the answer, 6. The digit 7 divides into 42 exactly 6 times.
Answer $42 \div 7 = 6$

Worked examples

Multiplication

1. $\begin{array}{r} 2 \\ \times 2 \\ \hline 4 \end{array}$ 2. $\begin{array}{r} 5 \\ \times 5 \\ \hline 25 \end{array}$ 3. $\begin{array}{r} 9 \\ \times 7 \\ \hline 63 \end{array}$

To multiply the above you really need to know your multiplication tables but here are some hints that will be of use.
When multiplying by the number 10, we can simply add a 0 to the number being multiplied:

6 x 10 = 60, 66 x 10 = 660 and 666 x 10 = 6660

If you multiply by 100 then add two 0s and by 1000, three 0s:

6 x 10 = 60, 6 x 100 = 600 and 6 x 1000 = 6000

When using a decimal point in a calculation you ignore it until you find the answer. Then count all the digits to the right of the points and place the decimal point in the answer by counting that number of digits from right to left

4. 33.44 x 3.2

$$\begin{array}{rl} 33.44 & \text{(two decimal places)} \\ 3.2 & \text{(one decimal place)} \\ \hline 107.008 & \text{(three decimal places)} \end{array}$$

5. 6.3 x 55.510

$$\begin{array}{rl} 55.510 & \text{(three decimal places)} \\ 6.3 & \text{(one decimal place)} \\ \hline 349.7130 & \text{(four decimal places)} \end{array}$$

Although it is not really important how you set out your sum, because we forget the decimal point, it is better to set the sum as shown in these examples as it reduces the chances of making a mistake.

6. 324.324 x 7654.34

$$\begin{array}{rl} 324.324 & \text{(3 dec.pl.)} \\ \times 7654.34 & \text{(2 dec.pl.)} \\ \hline 1297296 & \\ 9729720 & \\ 129729600 & \\ 1621620000 & \\ 19459440000 & \\ 227026800000 & \\ \hline 248248616616 & \\ \hline \end{array} = 5 \text{ decimal places.}$$

248248616616 Now count 5 decimal places from right to left

Answer 2 482 486.16616

Examples for practice:

7. 65.9 x 3.0   8. 2.33 x 17.1
9. 0.76 x 0.99   10. 7.00 x 0.55

Division

1. $2\overline{)22}$ = 11   2. $9\overline{)63}$ = 7   3. $6\overline{)48}$ = 8

Once again the multiplication and division table will help you in working out the above examples.

4. $16\overline{)346}$ = 21.6

For multiplication we used short cuts to help us speed up our work, and we can do the same with the division too. When dividing by 10 we can reduce the value of the given figure by moving a decimal point from right to left, as follows:

5. $10\overline{)654}$

Remove the last figure, and place it behind a decimal point, for example:
Answer 65.4

The same method can be used when we divide by 100 or 1000.

6. 654 ÷ 100

Remove the last two digits and place in a decimal point
Answer 6.54

When we divide with numbers containing a decimal point, the decimal point has to be removed first by multiplying by 10 or 100 or 1000 etc. depending upon the number of figures after the decimal point, and add an equivalent number of noughts to the figure being divided.

7. 2435 ÷ 3.6

$3.6\overline{)2435}$

$= 36\overline{)24350}$

A decimal point is then placed, following the nought, and a decimal point is then placed in the answer line immediately above as shown:

$$= 36\overline{)24350.}$$

```
       676.38
 = 36)24350.
      216
       275
       252
        230
        216
         140
         108
          320
          288
           32.
```

Answer 676.38

8. 87.54 ÷ 6
   = 6)87.54

Decimal point is placed in answer line

```
       .
   6)87.54
     14.59
 = 6)87.54
     6
     27
     24
      35
      30
       54
       54
       00
```

Answer 14.59

9. 72 ÷ 0.06
   = 0.06)72

The decimal point is moved two places then two noughts are added to the numerator, followed by a decimal point:

```
         .
 = 6)7200.
```

```
     1200.
= 6)7200.
    6
    12
    12
     00
     00
```

Answer 1200

Revision questions

Divide the following:

1. $10 \div 5$	2. $15 \div 3$	3. $18 \div 2$	4. $21 \div 7$	5. $45 \div 15$
6. $10 \div 2.5$	7. $13 \div 2$	8. $20 \div 2.1$	9. $13.6 \div 3.2$	10. $12.07 \div 4$

5 Decimals and fractions

The metric system uses decimals to represent portions of whole numbers. The imperial system uses *fractions* instead.

A fraction, ¾ for instance, is made up of two separate numbers. The lower number or *denominator* indicates how many equal parts the whole number has been divided into, in this case 4. The upper number, or *numerator*, indicates how many of these equal parts are present, in this case 3.

To convert a fraction to a decimal, divide the *denominator* (the figure below the line) into the *numerator* (the figure above the line).

Worked examples

1. Convert ¾ to a decimal.

$$\begin{array}{r} .75 \\ \tfrac{3}{4} = 4\overline{)3.00} \\ \underline{28} \\ 20 \\ \underline{20} \\ 00 \end{array}$$

Answer 0.75

2. Convert $^7/_8$ to a decimal.

$$\begin{array}{r} .875 \\ {}^7/_8 = 8\overline{)7.00} \\ \underline{64} \\ 60 \\ \underline{56} \\ 40 \\ \underline{40} \\ 00 \end{array}$$

Answer 0.875

3. Convert $^1/_{16}$ to a decimal.

$$
\begin{array}{r}
.0625 \\
^1/_{16} = 16\overline{)1.0000} \\
\underline{96} \\
40 \\
\underline{32} \\
80 \\
\underline{80} \\
00
\end{array}
$$

Answer 0.0625

4. Convert $^5/_8$ to a decimal.

$$
\begin{array}{r}
.625 \\
^5/_8 = 8\overline{)5.00} \\
\underline{48} \\
20 \\
\underline{16} \\
40 \\
\underline{40} \\
00
\end{array}
$$

Answer 0.625

5. Convert $^3/_8$ to a decimal.

$$
\begin{array}{r}
.375 \\
^3/_8 = 8\overline{)3.00} \\
\underline{24} \\
60 \\
\underline{56} \\
40 \\
\underline{40} \\
00
\end{array}
$$

Answer 0.375

Revision questions

Convert these fractions to decimals.

1. 1/16 2. 1/32 3. 1/8 4. 3/5 5. 3/16
6. 11/64 7. 7/8 8. 3/32 9. 5/16 10. 1/4

6 Ratios

A ratio is another method of presenting the information in a fraction.

Take the ratio 1:3. This means that for every 1 of the first there are 3 of the second, i.e. that there is $^1/_3$ as much of the first as the second.

The colon (:) in a ratio, therefore, separates the numerator and the denominator of a fraction, in that order:

1. $^7/_8$ = 7:8 ratio.
2. $2^1/_4$ = 9/4 = 9:4 ratio.

Worked examples

A concrete mixer contains 5 shovels of sand and 20 shovels of gravel. Find the ratio of sand to gravel.

Quantity of sand = 5 shovels
Quantity of gravel = 20 shovels
Therefore sand and gravel ratio = 5:20
Ratio = 5:20 or 1:4

In a mix of 24 shovels of sand and gravel, a ratio of 3:5 was used. Find the number of shovels of gravel used.

Ratio of sand and gravel = 3:5
In a single mix 3 + 5 shovels of sand and gravel are used.
In a mix of 24 shovels of sand and gravel $\frac{24}{8}$ are used, or 3 small mixes.

Therefore the ratio of sand and gravel equals:

$$3 \times 3 : 5 \times 3$$
$$= 9 : 15$$
$$= 3 : 5$$

Answer Number of shovels of gravel used = 15

Revision questions

Convert the following as ratios

1. 1/16 2. 1/4 3. 3/4 4. 7/8 5. 11/16 6. 23/4
7. 71/7 8. 35/6
9. Find the cement to aggregate ratio of a concrete mix if 8 portions of cement and 48 portions of aggregate are used.
10. In a mixture of solder, 50 parts of tin and lead are mixed to a ratio of 3:7. Find how many parts of solder are used.

7 Percentages

Yet another way of expressing quantities is by means of *percentages*. 'Percentage' means 'per hundred', and a percentage indicates how many $\frac{1}{100}$ parts are present. The sign used is %.

$$1\% = 1 \text{ in every } 100 = \frac{1}{100}$$

$$25\% = 25 \text{ in every } 100 = \frac{25}{100} = \frac{1}{4}$$

$$50\% = 50 \text{ in every } 100 = \frac{50}{100} = \frac{1}{2}$$

$$75\% = 75 \text{ in every } 100 = \frac{75}{100} = \frac{3}{4}$$

Other percentages, such as 21% or 34.5%, are less easy to express as fractions.

Percentages are used in calculating costs and prices in plumbing.

1. A plumber's paraffin blowlamp was originally priced at £6.82, to which an increase of 11% is added. Find the new price of the blowlamp.

New price = original price + percentage increase

$$\text{Percentage increase} = \text{original price} \times \frac{11}{100}$$

$$= 6.82 \times \frac{11}{100}$$

$$= 0.0682 \times 11$$

$$= £0.75$$

Since £1 = 100p, £0.75 = 75p

New price = original price + increase

= £6.82 + 75p

Answer New price = £7.57

A decrease in cost can be found using the same formula, then subtracting from the actual price instead of adding.

2. A wash handbasin with a minute chip underneath it is sold to a plumber. The salesman deducts 5% from the usual price, which is £4.64, because of the defect. Find the price of the basin.

   New price = original price – percentage decrease

   Percentage decrease = original price $\times \frac{5}{100}$

   $= 4.64 \times \frac{5}{100}$

   $= 0.0464 \times 5$

   = £0.23

   = 23p

   new price = original price – decrease

   = £4.64 – 23p

Answer New price = £4.41

There can be several different percentages involved at the same time.

3. Suppose a plumber buys a 32 mm tee. These used to have a *retail* price of £1.20.

   Now, as costs have increased, the manufacturer has raised all his retail prices by 30%.

   New retail price = old price + price increase.

   Price increase = old price x percentage increase.

   $= £1.20 \times \frac{30}{100}$

   = £0.36

   = 36p

   New retail price = £1.20 + 36p

   = £1.56

   The retail price is the one paid by the general public. The plumber can buy at a cheaper *trade* price. Suppose trade prices are 10% less than retail.

   Then: Trade price = retail price – trade discount.

   Trade discount = retail price x percentage discount

   $= £1.50 \times \frac{10}{100}$

   = £0.15

   = 15p

   Trade price = £1.50 – 15p

   = £1.35p

Finally we must allow for *Value Added Tax*, which is calculated as a percentage (currently 8%) and added on.

Price with VAT = price + VAT

VAT = price x percentage VAT rate

$= £1.35 \times \frac{8}{100}$

= £0.108

= 10.8p

= 11p to the nearest ½p

Price with VAT = £1.35 + 11p

= £1.46

4. Find the cost of a pair of straight-bladed tin snips costing £1.50 excluding VAT.

$\text{VAT} = £1.50 \times \frac{8}{100}$

= £0.120

= 12p

Price of snips plus VAT = £1.50 + 12p

Answer = £1.62

A quick way of finding 8% of £1.50 is this:

8% of £1.00 = 8p

8% of 50p = 4p

8% of £1.50 = 8p + 4p = 12p

Sometimes the percentages given are not the percentages by which the figure is increased or decreased, but the percentage left *after* the increase or decrease is allowed for.

So 10% trade discount can be put another way as:

Trade price = 100 – 10 = 90% of retail price.

There are some short cuts that can be taken. One such short cut is when adding or subtracting 10%.

5. Increase a price of £56.43 by 10%.

Place a row of digits, identical to the original figure, directly beneath but one decimal point to the right.

This gives you 10%, as $10\% = \frac{10}{100} = \frac{1}{10}$ of the original.

$$\begin{array}{r} = \quad 56.43 \\ + \quad 5.643 \\ \hline 62.073 \end{array}$$

Answer New price = £62.07

6. Reduce a price of £56.43 by 10%.
   We do the same, but subtract, rather than add.

$$\begin{array}{r} = \quad 56.43 \\ - \quad 5.643 \\ \hline 50.787 \end{array}$$

Answer New price = £50.79

We can do much the same for 100%, but this time we simply repeat the figure given. It is immediately clear that adding 100% to a figure doubles it. Subtracting 100% means taking itself away, that is, reducing it to zero.

7. Increase a price of £56.43 by 100%.

$$\begin{array}{r} = \quad 56.43 \\ + \, 56.43 \\ \hline 112.86 \end{array}$$

Answer New price = £112.86

8. Reduce a price of £56.43 by 100%.

$$\begin{array}{r} = \quad 56.43 \\ - \, 56.43 \\ \hline 00.00 \end{array}$$

Answer £00.00

To add 25% (¼), 50% (½) or 75% (¾) is fairly simple because we can generally calculate those fractions in our heads.

9. Reduce a price of £6.00 by 25%

$$\frac{\overset{1}{\cancel{25}}}{\underset{4}{\cancel{100}}} \times \frac{\overset{1.50}{\cancel{£6.00}}}{1} = £1.50$$

New price: £6.00 – £1.50 = £4.50

Answer £4.50

10. Increase a price of £12.00 by 50%

$$\frac{\overset{1}{\cancel{50}}}{\underset{2}{\cancel{100}}} \times \frac{\overset{6.00}{\cancel{£12.00}}}{1} = £6.00$$

New price: £12.00 + £6.00 = £18

Answer £18.00

Percentages are not used only for prices. Alloys used in the plumbing field consist of different parts, or percentages, of different metals.

11. A 0.454 kg bar of solder contains 7 parts of lead, 2.93 parts of tin and 0.07 parts of antimony. Express each part as a percentage.
Find the total number of parts = 7 + 2.93 + 0.07 = 10
To find the proportion of metal as a decimal we must calculate:

$$\frac{\text{Number of parts of metal}}{\text{Total number of parts}}$$

To convert this to a percentage, we must multiply by 100.

$$\text{So percentage lead} = \frac{\text{number of lead parts}}{\text{total number of parts}} \times 100$$
$$= \frac{7}{10} \times 100$$
$$= \frac{7}{1} \times 10$$

Answer 70% lead

$$\text{Percentage tin} = \frac{\text{number of tin parts}}{\text{total number of parts}} \times 100$$
$$= \frac{2.93}{10} \times 100$$
$$= \frac{2.93}{1} \times 10$$

Answer 29.3% tin

$$\text{Percentage antimony} = \frac{\text{number of antimony parts}}{\text{total number of parts}} \times 100$$
$$= \frac{0.07}{10} \times 100$$
$$= 0.07 \times 10$$

Answer 0.7% antimony

We can check the arithmetic by making sure these add up to the full 100%.

$$70 + 29.3 + 0.7 = 70 + 30$$
$$100$$

Revision questions

Convert the following fractions to percentages:

1. 1/100 2. 10/100 3. 1/8 4. 1/5 5. 1½
6. Add 10% to £12.00
7. Reduce £6.90 by 15%
8. Reduce £9.00 by 10%
9. Increase £15.00 by 3%
10. Increase £37.60 by 13%

8 Averages

To find an average, such as the average wage in a factory, average height of a football team or average speed of a car, we add together the figures in each instance and then divide by the number of instances.

Worked examples

1. Find the average height of this football team:

Heights of each player:	1.982 m	(6′6″)
	1.600 m	(5′3″)
	1.677 m	(5′6″)
	1.829 m	(6′0″)
	1.753 m	(5′9″)
	1.727 m	(5′8″)
	1.829 m	(6′0″)
	1.727 m	(5′8″)
	1.600 m	(5′3″)
	1.854 m	(6′1″)
	1.702 m	(5′7″)
Added heights are:	19.280 m	(63′3″)

Divide by 11

```
      1.752
  11)19.280
     11
      82
      77
       58
       55
        30
        22
         8
```

Answer Average height = 1.752 m (5′9″)

2. Find the average length of pipe, when the pieces measure

	153 mm	(6″)
	76 mm	(3″)
	457 mm	(18″)
	330 mm	(13″)
Added lengths are:	1016 mm	(3′4″)

Divide by 4

```
   254
4)1016
   8
   21
   20
    16
    16
    00
```

Answer Average length = 254 mm (10″)

Revision questions

Find the average of the following examples.

1. 4, 6, 8, 9, 32, 64, 18, 7, 11, 3
2. 2.5, 3.7, 6.4, 20, 17.1, 8
3. 0.62, 7, 0.03, 4.1, 0.1, 0.008, 0.3, 2.01

9 The use of symbols

A *symbol* is a sign used to indicate a relation or operation. In mathematics we use symbols to indicate relations:

$\times$ for multiplication, + for addition, – for subtraction, $\div$ for division, : for ratio.

Algebra uses letters instead of digits, so that we can work out general relations.

For instance, the area of a rectangle is given by the formula:

Area = length $\times$ breadth

if we indicate area by a

length by l

breadth by b

then we can say

$a = l \times b$

Multiplication in algebra can be abbreviated by missing out the $\times$ sign:

so $l \times b$ can be written lb

since $a = l \times b$ if the length of a field is 200 metres and the breadth 45 metres, then $l = 200$

$$b = 45$$
$$a = lb$$
$$a = 200 \times 45$$
$$a = 9000$$

Answer Field area = 9000 square metres

Another formula is for the perimeter (p) of a rectangle:

$$\begin{aligned} p &= 2l + 2b \\ &= 2 \times 200 + 2 \times 45 \\ &= 400 + 90 \\ &= 490 \end{aligned}$$

Answer Field perimeter = 490 metres

More complex formulae may use *powers* of numbers – squares, cubes etc. For instance, the area of a circle is given by

$a = \pi r^2$

Where a = area

π (pronounced pi) = a constant equal to about 3.142

r = radius

Suppose the radius of a circle is 6 metres then

$$\begin{aligned} r &= 6 \\ a &= \pi r^2 \\ &= 3.142 \times 6^2 \\ &= 3.142 \times 6 \times 6 \\ &= 3.142 \times 36 \\ &= 113.112 \end{aligned}$$

Answer Area = 113.112 square metres

When a group of factors such as $8ab$ is used, they are called a *term*, and each factor is a *coefficient* of the remainder of the term. 8 is the numerical coefficient of ab, and $8a$ is the coefficient of b. Note that the *numerical* coefficient is written before the *algebraic* coefficient.

There can be more than one term used in an expression, each term being separated by a plus sign or similar.

In an expression such as $8Z + 6Z^2 + 4Z^3$, 8 is the coefficient of Z, 6 is the coefficient of Z^2, and 4 is the coefficient of Z^3.

The coefficient then, is a number or other factor placed before another, as a multiplier.

In hydraulics, formulae are used in which symbols are used in their workings. One such formula, used to calculate the flow of water through a pipe is Box's formula, expressed as:

$$q = \frac{1}{5}\sqrt{\frac{d^5 H}{L}}$$

Where H = head in metres

d = diameter in centimetres

L = length in metres

q = quantities in litres/sec.

This formula includes the sign $\sqrt{\quad}$, which indicates the *square root* of the number. The square root of a number is the figure which when squared (multiplied by itself) will exactly equal that number.

For instance, $5^2 = 5 \times 5 = 25$

So $\sqrt{25} = 5$

In this example

$H = 8$ m
$d = 20$ mm
$L = 16$ m
q = unknown to be found.

NB. The formulae specifies d in cm, so we must convert

1 cm = 10 mm
$d = 2$ cm.

First we substitute the figures into the formulae.

$$q = \frac{1}{5}\sqrt{\frac{(2)^5 \times 8}{16}}.$$

Then we calculate q step by step. We begin at the buried centre of the complex sum, that is with the brackets.

$$q = \frac{1}{5}\sqrt{\frac{32 \times 8}{16}}.$$

Then we multiply within the square root above the line.

$$q = \frac{1}{5}\sqrt{\frac{256}{16}}$$

Then we divide within the square root.

$$q = \frac{1}{5}\sqrt{16}$$

Now since $4^2 = 4 \times 4 = 16$, the square root of 16 is 4, so

$$q = \frac{1}{5} \times 4$$

$$q = 0.8$$

Answer Flow 0.8 litres per second

This example used 16, one of the few numbers with a square root that is a whole number. Generally, square roots of numbers are found by using tables.

The same sign as is used for square roots can be used for greater powers when that power is displayed:

$\sqrt[3]{27}$ means the cube root of 27, that is the number which when cubed makes 27 ($3 \times 3 \times 3 = 27$).

10 Formulae

Perimeter of *square*	L = length of one side $= 4L$
Perimeter of *rectangle*	L = length long side b = length short side $= 2L + 2b$
Perimeter of *triangle*	a = length side 1 b = length side 2 c = length side 3 $= a + b + c$
Circumference of *circle*	d = diameter of circle $= \pi d$
Area of *square*	L = length of one side $= L^2$
Area of *rectangle*	L = length long side b = length short side $= lb$
Area of *triangle*	b = length of base h = height of triangle $= \frac{b \times h}{2}$
Area of *circle*	r = radius of circle $= \pi r^2$
Area of *circle*	d = diameter of circle $= \frac{\pi d^2}{4}$
Surface area of open-ended *cylinder*	r = radius of cylinder h = height of cylinder $= 2\pi rh$

also	d = diameter of cylinder
	$= 2r$
	h = height of cylinder
	$= \pi dh$
Surface area of open-ended *cone*	r = radius of cone
	l = slant height
	$= \pi rl$
Surface area of *sphere*	d = diameter of sphere
	$= \pi d^2$
Area of *trapezium*	b = length base
	t = length top
	h = height
	$= (b + t) \times \frac{h}{2}$
Volume of *cube*	l = length of one side
	$= l^3$
Volume of *rectangular block* (*cuboid*)	l = length
	b = breadth
	d = depth
	$= l \times b \times d$
Volume of *cylinder*	r = radius of cylinder
	h = height of cylinder
	since $r = d/2$
	d = diameter of cylinder
	h = height of cylinder
	$= \pi r^2 h$
	$= \frac{\pi d^2 h}{4}$
Volume of *cone*	r = radius of cone
	h = perpendicular height
	$= \frac{\pi r^2 h}{3}$
	d = diameter of cone
	since $r = d/2$
	$= \frac{\pi d^2 h}{12}$

Volume of *sphere*

r = radius of sphere

$= \frac{4\pi r^3}{3}$

d = diameter of sphere

since $r = d/2$

$= \frac{\pi d^3}{6}$

Volume of *trapezoidal*

b = length of base

t = length of top

h = height of section

l = length of section

$= b + t \times \frac{h}{2} \times l$

If we measure lengths in metres and calculate volumes, then the volume is in cubic metres. Generally we prefer to measure volume in litres. Since 1 cubic metre = 1000 litres, we can find capacity in litres by multiplying volume by 1000.

Capacity of *cube* (litres) $= L^3 \times 1000$

Capacity of *rectangular block* or *cuboid* (litres) $= l \times b \times d \times 1000$

Capacity of *cylinder* (litres) $= \pi r^2 h \times 1000$

Capacity of *cone* (litres) $= \frac{\pi r^2 h}{3} \times 1000$

Capacity of *sphere* (litres) $= \frac{\pi d^3}{6} \times 1000$

Capacity of *trapezoidal* (litres) $= b + t \times \frac{h}{2} \times l \times 1000$

or $\left(\frac{b+t}{2}\right) hl \times 1000$

Example

What is the cubic capacity of a tank in the shape of a cube, each side 600 mm long?

Capacity of cube (litres) $= L^3 \times 1000$

Where L is length of side in metres

Since 1000 mm = 1 m

600 mm long?

Capacity $= 0.6^3 \times 1000$
$= 0.216 \times 1000$
$= 216$ litres

Answer Capacity = 216 litres

Rearranging formulae

Most problems can be solved by using the right formula. Sometimes the same formula is used to solve several problems, but the formula is not always in the form required. When this happens the formula has to be rearranged to suit.

In a formula there is always an equal sign, and this equal sign plays an important part when rearranging formulae. When a digit or symbol is moved from one side of an equal sign to the other, the role that the digit or symbol is playing is reversed.

A *plus* sign becomes a *minus* sign:

$$a + 6 = b$$

is the same as

$$a = b - 6$$

Conversely a *minus* sign becomes a *plus* sign:

$$x - 8 = y$$

is the same as

$$x = y + 8$$

A *multiplication* sign becomes a *division* sign:

$$d \times 4 = f$$

is the same as

$$d = f \div 4$$

Conversely a *division* sign becomes a *multiplication* sign:

$$p \div 9 = q$$

is the same as

$$p = q \times 9$$

Very often multiplications and divisions are expressed using the other notation which does without multiplication and division signs:

$$4d = f$$

and

$$\frac{p}{9} = q$$

In this case we use cross-multiplication:

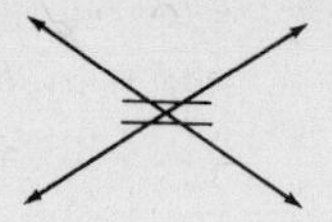

So a figure on the *top* line moves to the *bottom* line when carried to the other side of the = sign:

$$4d = f$$

is the same as

$$d = \frac{f}{4}$$

Conversely a figure on the *bottom* line moves to the *top* when carried to the other side of the = sign:

$$\frac{p}{19} = q$$

is the same as

$$p = 19q$$

A more complex example, with both top and bottom figures is:

$$\frac{7t}{u} = w$$

is the same as

$7t = uw$ (*u* changes sides and goes from bottom to top)

is the same as

$t = \frac{uw}{7}$ (7 changes sides and goes from top to bottom)

In powers and roots too we have this reversal taking place. A square on one side can be turned into a square root on the other.

So $r^2 = AB$

is the same as

$$r = \sqrt{AB}$$

Conversely

$$p = \sqrt{CD}$$

is the same as

$$p^2 = CD$$

is the same as

$$\frac{p^2}{C} = D$$

In the same equation we may move multiples also, using the cross-multiplication rules.

In these examples, all three sets of rules (for add/subtract, multiply/divide and square/square root) are used.

1. The formula for the capacity of a cylindrical tank (page 35) is:

$$C = \pi r^2 h 1000$$

where C = capacity in litres
r = radius
h = height

2. Suppose we know capacity and radius and wish to have the height. Then we must rearrange the formula to make h its subject.

$$C = \pi r^2 h 1000$$

by the cross-multiplication rule

$$\frac{C}{\pi r^2 1000} = h$$

As the subject of an equation is usually written on its left side, this is rewritten as:

$$h = \frac{C}{\pi r^2 1000}$$

3. Suppose we know capacity and height and wish to have the radius. Then we must rearrange the formula to make r its subject:

$$C = \pi r^2 h 1000$$

$$r^2 = \frac{C}{\pi h 1000}$$

Then we must square root both sides:

$$r = \sqrt{\frac{C}{\pi h 1000}}$$

Revision questions

Rearrange the following

1. $\frac{A}{B} = C$ to $A =$
2. $A - B = C$ to $B =$
3. $AB = CD$ to $C =$
4. $\frac{AB}{D} = C$ to $D =$
5. $6 + 2 = 8$ to $6 =$
6. $A = \sqrt{D/4}$ to $D =$
7. $\sqrt{A} = CD$ to $C =$
8. $6 = \sqrt{4/A}$ to $A =$
9. $q = \sqrt{3D^5 H/L}$ to $H =$
10. $q = \frac{\sqrt{3D^5 H}}{L}$ to $L =$

11 Pythagoras' theorem and square root tables

Pythagoras was a Greek philosopher who lived about 2500 years ago. He devised a way of calculating the length of one side of a right angle triangle when the length of the other two sides is known. The method is called Pythagoras' theorem.

His formula is: 'In a right angle triangle, the square on the hypotenuse is equal to the sum of the squares on the sides forming the right angle.'

The *hypotenuse* is the longest side in the right angle triangle. It is always opposite the right angle.

Let us call the length of the hypotenuse x
the length of one short side b
the length of the other short side h

Then the square on the hypotenuse is x^2
and the squares on the other two sides are b^2 and h^2
so the sum of the squares on the other two sides is $b^2 + h^2$
As a formula:

$$x^2 = b^2 + h^2$$

Taking the example shown we have $b = 3, h = 4$

$$x^2 = b^2 + h^2$$
$$x^2 = 3^2 + 4^2$$
$$x^2 = 9 + 16$$
$$x^2 = 25$$
$$x = \sqrt{25}$$

Answer $x = 5$

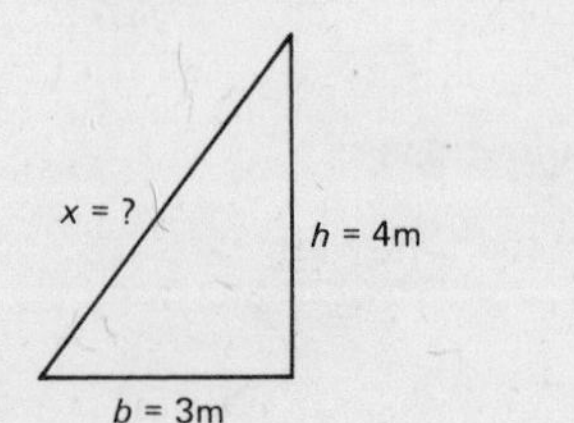

Fig. 1

So this right angled triangle has sides 3, 4 and 5 metres long. Conversely, if we draw a triangle with sides 3, 4 and 5 metres long we know it must have a right angle in it. This is handy when a right angle is wanted on a building site. A 3–4–5 triangle can be quickly put together with tapes 3 metres, 4 metres and 5 metres long to produce an accurate right angle.

Other problems can be solved with the theorem of Pythagoras. We shall look at a few of them.

Worked examples

1. If the hypotenuse length is given along with the length of one side, then we can find the remaining side as follows:

Using Pythagoras' theorem:

$h^2 = x^2 - b^2$
$h^2 = 13^2 - 5^2$
$h^2 = 169 - 25$
$h^2 = 144$
$h = \sqrt{144}$

Answer $h = 12$ m

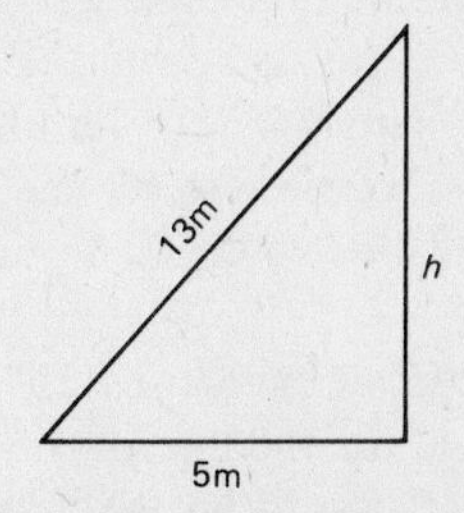

Fig. 2

2. To find length b on the piping sketch (Figure 3), using Pythagoras' theorem:

$b^2 = x^2 - h^2$
$b^2 = 10^2 - 8^2$
$b^2 = 100 - 64$
$b^2 = 36$
$b = \sqrt{36}$

Answer $b = 6$

16m
8m

Fig. 3

The three forms of the theorem we use are:

$x^2 = h^2 + b^2$: to find the length of the hypotenuse
$b^2 = x^2 - h^2$: to find the length of the base
$h^2 = x^2 - b^2$: to find the length of the height

Revision questions

1. Find the length of hypotenuse of a triangle having a base length of 5 m and a vertical height of 6 m.
2. A triangle has a base length of 3 m and an hypotenuse length of 6 m. What is the vertical height?
3. What is the length of base of a triangle having an hypotenuse length of 8 m and a vertical height of 5 m?

Square root tables

The tables are printed on four pages usually as part of a four figure tables book. The first page of the square root tables reads from 10 to 32 down the left hand column vertically and the following three pages 33 to 54, 55 to 77 and 78 to 99 respectively. The top line of digits on all four pages read from 0 to 9 followed by a smaller row of digits reading from 1 to 9.

There are no tables for single numbers such as the number 4 down the vertical left hand columns, so look for the number 40 on page two. Now move to the right under the zero column found on the top of the page. In this column there are two different rows of digits to choose from, 2000 and 6325 underneath.

To choose the correct number use the rule given below:

For number 4 use the top row of digits,
For number 40 use the bottom row of digits,
For 400 use the top row, and
For 4 000 use the bottom row, and so on.

In this particular case the square root of 4 is required so from the rule the top row of digits is selected, 2000. It is obvious that 2000 is not the square root of 4 until the decimal point is positioned to give the correct answer.

To find the position of the decimal point.

Most four figure tables state on the bottom of the page that to find the first significant number and position of the decimal point is by inspection, but to help determine these positions Table 5 may be used for practice.

Table 5

Number	*Square Root*	*Line*
0.04	0.2	Top Line
0.4	0.6325	Bottom Line
4.0	2.0	Top Line
40.0	6.325	Bottom Line
400.0	20.0	Top Line
4 000.0	63.25	Bottom Line

Consider the numbers which have their square roots on the top line in the Number column, 0.04, 4, 400. In each case the decimal point in the number moves two places to the right, i.e. 0.04 to 4, and 4 to 400. but in the square root column of Table 5, the decimal point only moves one place to the right each time, i.e. 0.2 to 2, and 2 to 20.

From Table 5 given, the square root of 4 must be 2. after placing the decimal point in the correct place.

If the square root of 4 is 2 and this is remembered, it is easy enough after a little practice to determine the correct position of the decimal point in any answer.

If 2 is the square root of 4 then by moving the decimal point, in the square root column, one place to the right we have the square root of 400 (20) and by moving the decimal point to the left we have the square root of 0.04 (0.2).

The square roots of 0.4, 40 and 4 000 are found in the same way using the bottom row of digits.

Sometimes the square roots of numbers with decimal parts must be found, such as the square root of 6.45. In this case look for the number 64 in the left hand vertical column of the tables, third page, and move to the right until under the 5 column found along the top row of digits of the page. Here again are two rows of digits to choose from, choose the top row as explained previously, 2540. The decimal point must now be placed as previously shown.

The square root of 6.45 is 2.459.

It may be noted that there may be a slight difference in answers taken from four figure tables and five figure tables. This is because the last digit may be rounded off to the next whole number as shown below.

The square root of 59.7 is 7.7265775, by calculator, rounded off to 7.7266 in five figure table or 7.727 in four figure table.

Examples

Find the square roots of the following:

1. 5	2. 17	3. 6.3	4. 20.42
5. 625	6. 400	7. 49	8. 0.6
9. 93.8	10. 100.		

Revision questions

Find the square roots of the following using square root tables.

1. 5	2. 17	3. 6.3	4. 20.42	5. 625	6. 400
7. 49	8. 0.6	9. 93.8	10. 100		

Worked examples on chapters 10 and 11

Perimeter

1. Find the length of perimeter around a square field measuring 6.6 m along one side

Formula perimeter = $4L$ where L = length

= 4×6.6 m

Answer 26.4 m

2. Find the length of perimeter around a rectangular building measuring 4 m x 7 m

Formula = $2 \times a + 2 \times b$

= $2 \times 4 + 2 \times 7$

= 8 + 14

Answer 22 m

3. Find the length of perimeter around a triangular plot of land, measuring 3 m x 4 m x 5 m,

Formula = $a + b + c$

= $3 + 4 + 5$

Answer 12 m

Circumference

4. Find the circumference of a circle having a diameter of 9 m.

Formula = $\pi \times d$

= 3.142×9

Answer 28.278 m

Area

5. Find the area of a square, the length of one side being 8 m.

Formula = l^2

= 8^2

= 8×8

Answer 64 m^2

Area

6. Find the area of a school playing field which is rectangular in shape, measuring 12 m x 16 m.

Formula $= L \times B$

$= 16 \times 12$

Answer $= 192\ m^2$

7. A right angle triangle measuring 9 m at the base and 8 m in perpendicular height is to be grassed in a garden. Find the area to be grassed.

Formula $= \dfrac{b \times h}{2}$

$= \dfrac{9 \times 8}{2}$

$= \dfrac{72}{2}$

Answer $= 36\ m^2$

8. A circular swimming pool has a diameter of 10 m. Find its area.

Formula $= \pi \times r^2$

$= 3.142 \times 5 \times 5$

$= 3.142 \times 25$

Answer $= 78.55\ m^2$

9. Find the area of the surface sides of a cylinder having a perpendicular height of 4 m and a diameter of 2 m. Ignore the flat ends.

Formula $= 2\pi r\, h$

$= 2 \times 3.142 \times 1 \times 4$

$= 2 \times 3.142 \times 4$

$= 8 \times 3.142$

Answer $= 25.136\ m^2$

10. A cone shaped roof is to be covered with lead. Ignoring any allowance for rolls etc., calculate the area of lead required. The diameter is 10 m and the perpendicular height is 16 m.

Formula $= \pi r l$

$= 3.142 \times 5 \times l$

We do not know what the value of l is, this will have to be found by using Pythagoras' theorem, for finding the length of the hypotenuse of a right angle triangle. l is the hypotenuse.

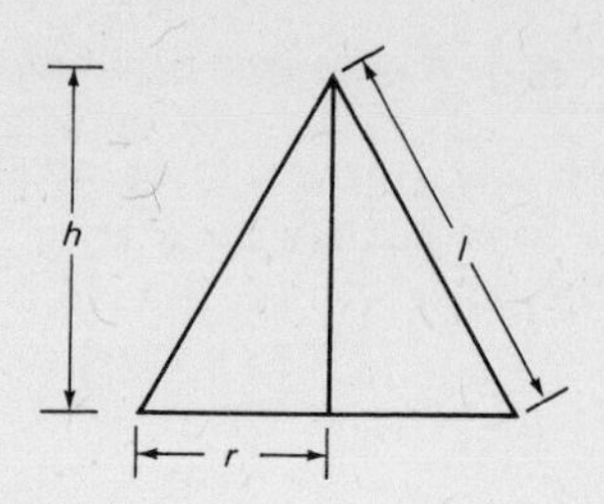

Therefore

$$l^2 = r^2 + h^2$$
$$l^2 = 5^2 + 16^2$$
$$l^2 = 25 + 256$$
$$l^2 = 281$$
$$l = \sqrt{281}$$
$$l = 16.76$$

Now we have the value of l (16.76) we can use the formulae given.

Formula $= \pi r l$

$= 3.142 \times 5 \times 16.76$

$= 3.142 \times 83.8$

Answer $263.3\ m^2$

11. Find the area of a sphere when the diameter is 8 m.

Formula $= \pi d^2$

$= 3.142 \times 8^2$

$= 3.142 \times 64$

Answer $201.1\ m^2$

12. Find the area of a trapezium, having a base of 8 m a top of 6 m and a height of 12 m.

Formula $= (b + t) \times \frac{h}{2}$

$= (8 + 6) \times 6$

$= 14 \times 6$

Answer $84\ m^2$

Volume

13. Find the volume of a cube, when the length of one side is 4 m.

Formula $= l^3$

$= 4 \times 4 \times 4$

$= 16 \times 4$

Answer $64\ m^3$

14. A rectangular galvanized storage cistern measures 4 m x 3 m x 2 m. Find the cubic volume of the cistern.

Formula $= l \times b \times d$

$= 4 \times 3 \times 2$

$= 12 \times 2$

Answer 24 m^3

15. What is the volume, in $metres^3$, of a cylinder measuring 2 m high by 1 m in diameter?

Formula $= \pi r^2 h$

$= 3.142 \times 0.5 \times 2.5 \times 2$

$= 3.142 \times 0.25 \times 2$

$= 3.142 \times 0.5$

Answer 1.571 m^3

16. A conical shaped hole has to be filled. Find the number of cubic metres required if the diameter of the cone is 14 m, and the depth is 9 m.

Formula $= \dfrac{\pi r^2 h}{3}$

$= \dfrac{3.142 \times 7 \times 7 \times 9}{3}$

$= \dfrac{3.142 \times 49 \times 9}{3}$

$= \dfrac{3.142 \times 441}{3}$

$= \dfrac{1385.62 \text{ m}^3}{3}$

Answer 461.87 m^3

17. Find the volume of a sphere, having a diameter of 10 m.

Formula $= \dfrac{4 \times \pi r^3}{3}$

$= \dfrac{4 \times 3.142 \times 5 \times 5 \times 5}{3}$

$= \dfrac{3.142 \times 500}{3}$

$= \dfrac{1571}{3}$

Answer 523.3 m^3

18. A retaining wall is to be made of concrete, the shape of which is trapezoidal in the end section and the length of the wall 60 m. The dimensions of the end sections are as follows: top = 2 m, base = 3 m, and the vertical height is 6 m. Find the number of cubic metres of concrete required to make the wall.

Formula $= \frac{(b + t)}{2} hl$

$= \frac{(3 + 2)}{2} \times 6 \times 60$

$= 2.5 \times 6 \times 60$

Answer 900 m^3

19. Calculate the depth of a cold water storage cistern when the width is 2 m, the length is 3 m, and the cistern capacity is 60,000 litres.

Formula $q = l \times b \times d \times 1000$

Rearrange formula to equal d

$d = \frac{q}{l \times b \times 1000}$

$= \frac{60\,000}{3 \times 2 \times 1000}$

$= \frac{60\,0\not{0}\not{0}\not{0}}{6\not{0}\not{0}\not{0}}$

$= \frac{60}{6}$

Answer 10

Check Using the answer found:

$q = l \times b \times d \times 1000$

$q = 3 \times 2 \times 10 \times 1000$

$q = 60 \times 1000$

$q = 60\,000$ litres

20. A cistern is square on plan and is 800 mm in depth. If the cistern will hold 288 litres when full, what is the length of one side?

Formula $q = l \times b \times d \times 1000$

Since it is square $l = b$ and we have $l \times b = l^2$

$q = l^2 \times d \times 1000$

$l^2 = \frac{q}{d \times 1000}$

$l^2 = \frac{288}{0.8 \times 1000}$

$$l^2 = \frac{288}{800}$$

$$l^2 = 0.36$$

$$l = \sqrt{0.36}$$

Answer $l = 0.6$ m (or 600 mm)

Check Using answer found:

$q = l^2 \times d \times 1000$

$q = 0.6^2 \times 0.8 \times 1000$

$q = 0.288 \times 1000$

$q = 288$ litres

12 Moments

The moment of a force can be defined as the tendency of a force to turn an object about a point. The point of turning is called the *fulcrum*. There are two main points to consider when assessing the state of equilibrium on a specific bar or beam, etc.

1. Force applied, expressed in newtons or kilograms force.
2. Distance from the fulcrum to the point at which the force is applied.

In the following examples we use both the force and the distance, and both newtons and kilograms force.
To convert kilograms force (kgf) to newtons (N), multiply kgf by 9.81.

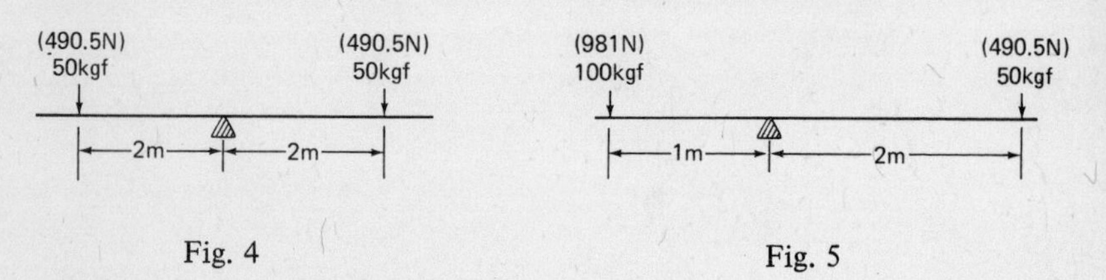

Fig. 4

Fig. 5

To maintain a state of equilibrium the anti-clockwise moments (ACWM) must equal the clockwise moments (CWM) as they do in Figures 4 and 5.

Figure 4

CWM = Force x distance		ACWM = Force x distance	
(kgf)	(N)	(kgf)	(N)
kgf x m	N x m	kgf x m	N x m
= 50 x 2	= 490.5 x 2	= 50 x 2	= 490.5 x 2
= 100 kgf – m	= 981 Nm	= 100 kgf – m	= 981 Nm

It can be seen that the CWM equal the ACWM, providing a state of equilibrium.

Figure 5

CWM = Force x distance		ACWM = Force x distance	
(kgf)	(N)	(kgf)	(N)
kgf x m	N x m	kgf x m	N x m
50 x 2	490.5 x 2	100 x 1	981 x 1
100 kgf	981 N/m	100 kgf	981 N/m

In this example too there is a state of equilibrium.
The following examples will be worked out in kgf only but conversion to newtons can be obtained by using the conversion factor of
9.81 N = 1 kgf.

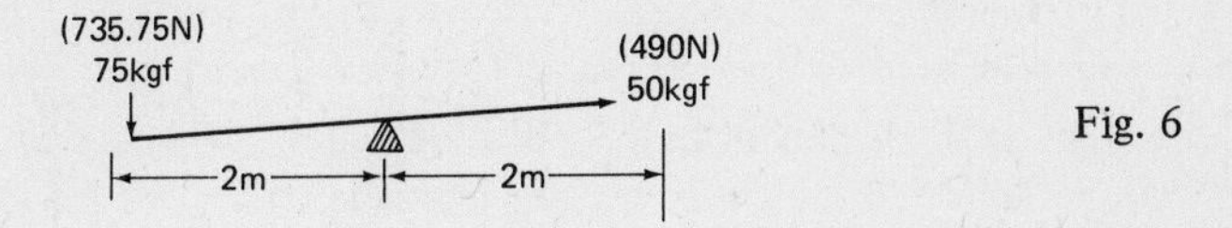

Fig. 6

Figure 6

CWM = force x distance	ACWM = force x distance
= kgf x m	= kgf x m
= 50 x 2	= 75 x 2
100 kgf – m	150 kgf – m

It can be seen that the CWM do not equal the ACWM and so the state of equilibrium is lost.

In the three examples shown, all the information was given, but this is not always the case, and so sometimes we have to find the missing dimension. When this happens we have to rearrange the given information to find the missing one.

If we consider Figure 7 we do not know the length of *B* to maintain a state of equilibrium, so we will have to rearrange the information given to find it, as follows.

75kgf (*A*)
50kgf (*C*)
? (*B*)
Fulcrum
2m (*D*)

Fig. 7

Information known: $AB = CD$
Rearranging the formulae to suit:

$$= B = \frac{CD}{A}$$

Substituting numbers for letters:

$$= B = \frac{50 \times 2}{75}$$

$$= B = \frac{100}{75}$$

Answer $B = 1.333$ m

On the other hand we may have to find '*A*' (Figure 7) to maintain a state of equilibrium. Assuming *B* to be 2 metres in length.
Information known: $AB = CD$
Rearranging the formulae to suit:

$$A = \frac{CD}{B}$$

Substituting numbers for letters:

$$A = \frac{50 \times 2}{2}$$

$$A = \frac{100}{2}$$

Answer $A = 50$ kgf

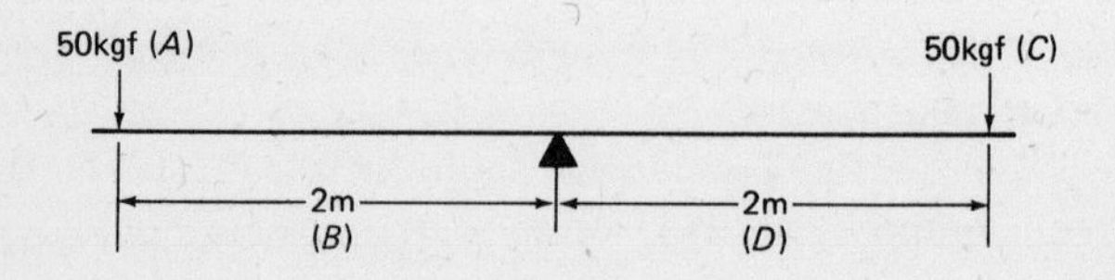

Fig. 8

Again if we substitute 50 kgf into Figure 8 we would have a state of equilibrium. We can check this as before.
Checking: *Fig. 8.*

CWM = force x distance
= kgf x m
= 50 x 2
= 100 kgf – m

ACWM = force x distance
= kgf x m
= 75 x 1.333
= 100 kgf – m (approx.)

Checking: *Fig. 8.*

CWM = force x distance	ACWM = force x distance
= kgf x m	= kgf x m
= 50 x 2	= 50 x 2
Answer 100 kgf – m	*Answer* 100 kgf – m

We now have a state of equilibrium.

In the field of plumbing we use the principles of moments, sometimes quite unknowingly, when we use a Stillson wrench or a spanner. So let us take a closer look at this operation, using a Stillson wrench on a steel pipe.

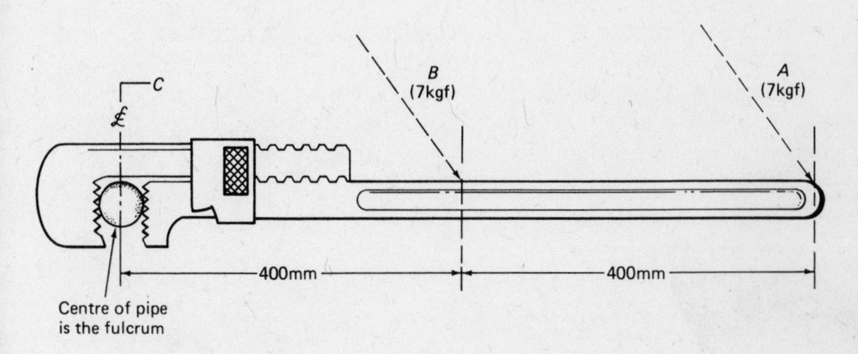

Fig. 9

It will be seen that the further we travel from the fulcrum, the more our effort is magnified, and the easier it becomes to turn the Stillson wrench on the pipe.

The moment of a force is the applied force (kgf) multiplied by the distance from the applied force to the fulcrum (centre of the pipe).

CWM at $C = 0$ (no turning moment, distance, at all).

CWM at B = force x distance.

B = kgf x m

B = 7 kgf x 0.4 m

Answer B = 2.8 kgf – m

CWM at A = force x distance

A = kgf x m

A = 7 kgf x 0.8

Answer A = 5.6 kgf – m

It is usual when calculating the magnitude or moments of a force, acting on a Stillson wrench or spanner, to express the reaction as *torque*, because *torque* = *force* x *radius*, and *torque* is expressed in *newtons.*

To convert kgf to newtons multiply by 9.81 (10 may be used for quick calculations). This is because there are 9.81 newtons to 1 kgf.

When cantilever brackets are used to support a wash-hand basin or a cold water storage cistern, we also use the principle of moments.

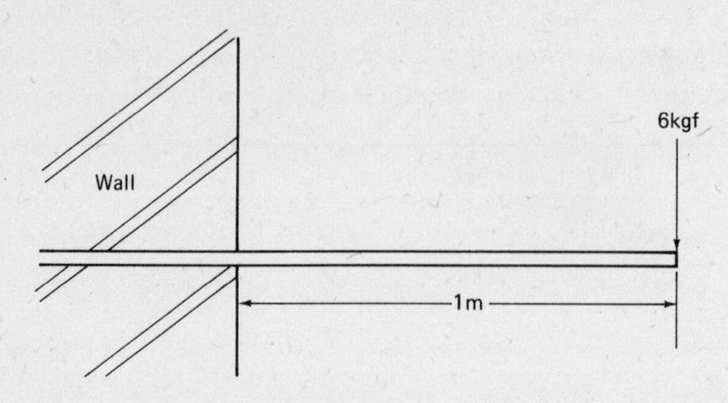

Fig. 10

The force acting on the lever would produce a turning effect. The moments of a force for the cantilever about the wall would be as follows.

CWM = force x distance
= kgf x m
= 6 kgf x 1 m

Answer 6 kgf – m

This means that the wall around the cantilever bracket would have to withstand the 6 kgf – m turning effect to maintain a state of equilibrium and so support the chosen cistern.

Lintels over doorways or windows are also based on the principles of moments just like the cantilever lintel (or step). In the final two examples we will look at each of these.

Calculate the distance value of D in Figure 11 if the system shown is to remain in balance in a horizontal position.

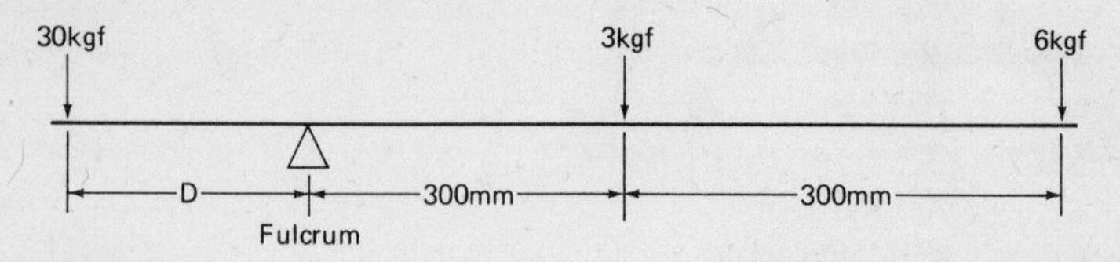

Fig. 11

CWM	= ACWM
(6 x 600) + (3 x 300)	= (30 x D)
(6 x 0.6) + (3 x 0.3)	= (D 30)
(3.6 + 0.9)	= (D 30)
4.5	= (D 30)

then

$$\frac{4.5}{30} = D$$

$$= 0.150 = D$$

Answer $D = 0.150$ mm

Calculate the reactions on the supports A and B created by the concrete beam in Figure 12.

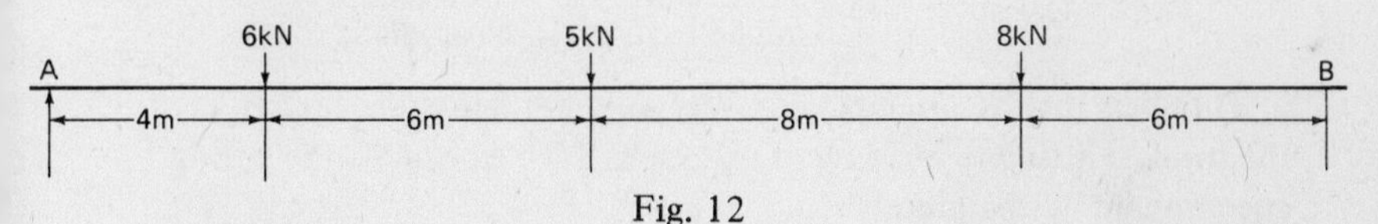

Fig. 12

CWM must equal ACWM

Ra 24 =	(8 x 6)	+	(14 x 5)	+	(20 x 6)	=	(4 x 6)	+	(10 x 5)	+	(18 x 8) 24 Rb
Ra 24 =	48	+	70	+	120	=	24	+	50	+	144 24 Rb
Ra 24 =			238			=			218		24 Rb

therefore:

$$Ra = \frac{238}{24} = \frac{218}{24} = 24\ Rb$$

$$Ra = 9.916 \text{ kN} = 9.083 \text{ kN} = Rb$$

Check

Total loading = 6 + 5 + 8 = 19 kN

Reactions at A and B = 9.916 + 9.083 = 18.999

Answer Say 19 kN

Revision questions

1. A Stillson wrench is used to turn a pipe. The wrench is 1 m long and forces of 5 kgf and 10 kgf are acting on the wrench at 500 mm and 1 m from the centre of the pipe (fulcrum) in a clockwise manner. Determine the moment of force at each.
2. Two cantilever brackets are built into a wall to support a cistern which exerts a force of 400 kgf on the free ends of them. Calculate the clockwise force on the wall surrounding one bracket. The brackets extend 800 mm from the wall.

13 Levers

Levers can be defined as devices used to magnify a force applied to move an object. This magnification is sometimes called mechanical advantage, and can be expressed as:

$$\text{Mechanical advantage} = \frac{\text{weight of load raised}}{\text{applied effort (force applied)}}$$

Plumbers use the principle of levers daily in their work, for instance in the form of lifting or bending of materials. Take an example of lifting encountered by the plumber.

A large heavy boiler has to be lifted at one end, so the plumber uses a length of pipe or timber to raise the end, probably pushing the pipe or timber down over a house brick. The pipe or timber is acting as the lever and the house brick as the fulcrum as shown in Figure 13

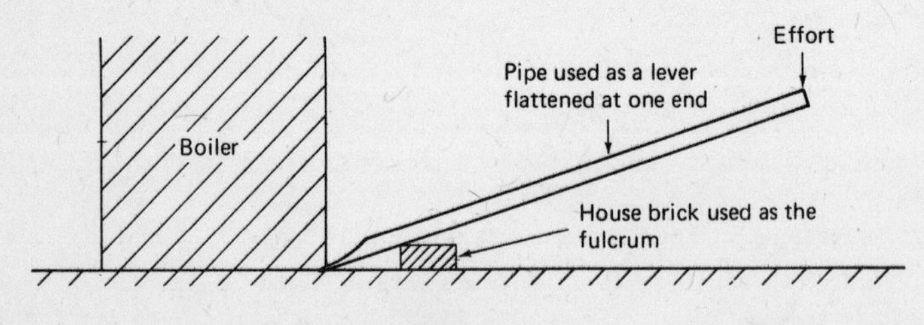

Fig. 13

For our calculations we can simplify the diagram as shown in Figure 14.

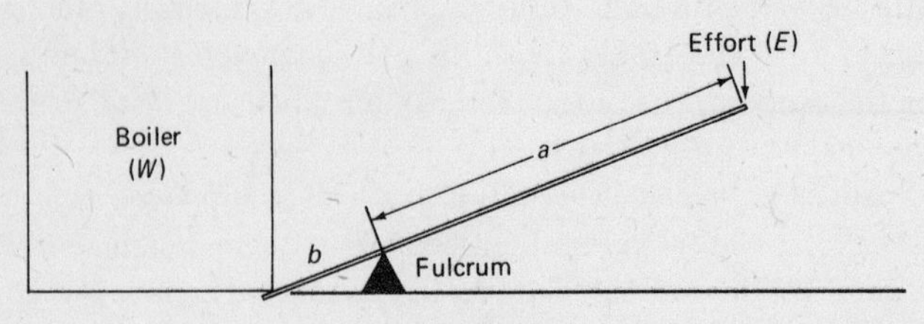

Fig. 14

The length of the bar sections a and b affect the ease of levering as does the amount of effort E and the weight of the load W. In the following examples we will calculate each of these in turn, when the others are known.

N.B. From previous notes on moments you will remember how to rearrange any formula to our advantage, remembering that the equal sign alters any powers when crossed.

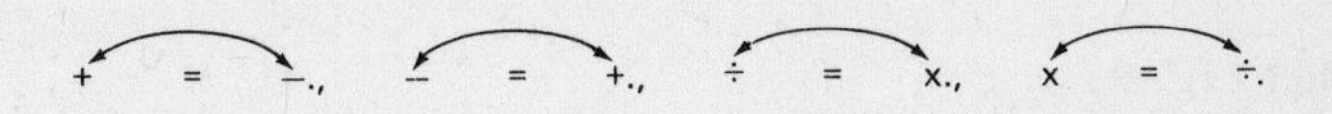

The formula used in levers is:

$$Wb = Ea$$

To find effort (E)

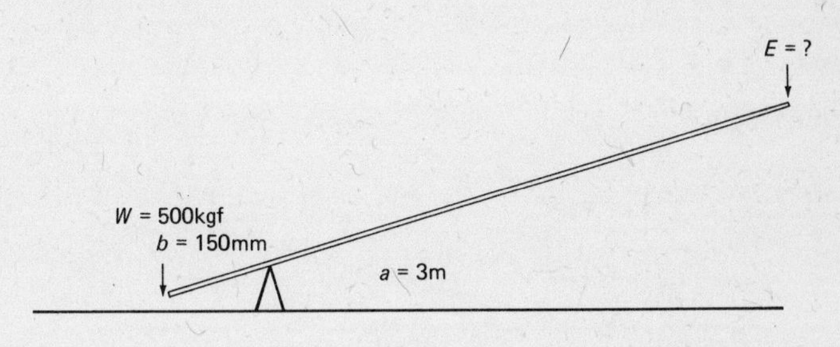

Fig. 15

Formula = $Wb = Ea$.
Rearranging formula to suit:

$$E = \frac{Wb}{a}$$

Substituting letters for numbers:

$$E = \frac{500 \times 0.15}{3}$$

$$E = \frac{75}{3}$$

Answer $E = \underline{25 \text{ kgf}}$

To find weight (W)

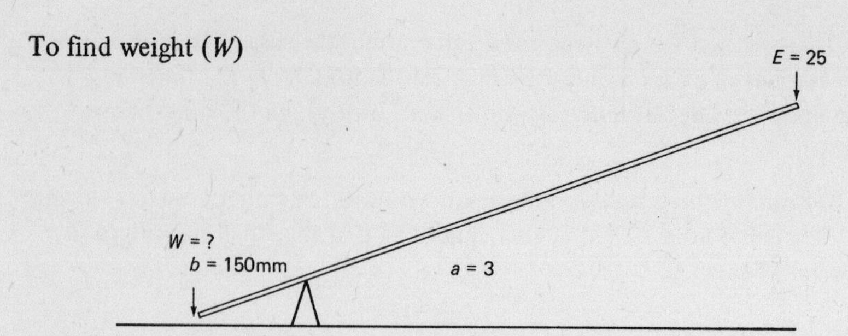

Fig. 16

Formula = $Wb = Ea$
Rearranging the formula to suit:

$$W = \frac{Ea}{b}$$

Substituting

$$W = \frac{25 \times 3}{0.15}$$

$$W = \frac{75}{0.15}$$

Answer $W = \underline{500 \text{ kgf}}$

Finding length b.

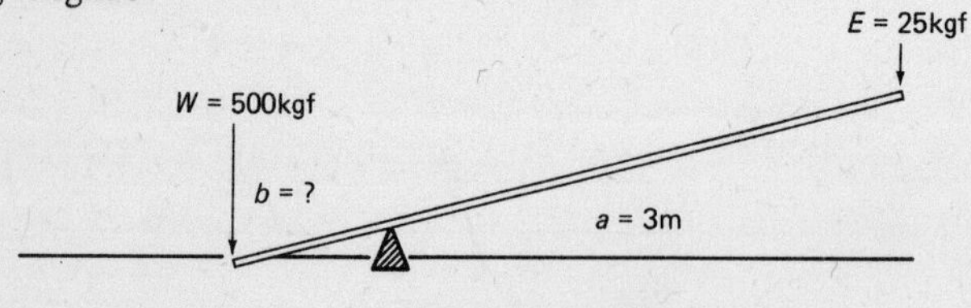

Fig. 17

Formula = $Wb = Ea$.
Rearranging formula to suit:

$$b = \frac{Ea}{W}$$

Substituting letters for numbers:

$$b = \frac{25 \times 3}{500}$$

$$b = \frac{75}{500}$$

$$b = 0.15 \text{ m}$$

Answer $b = 150 \text{ mm}$

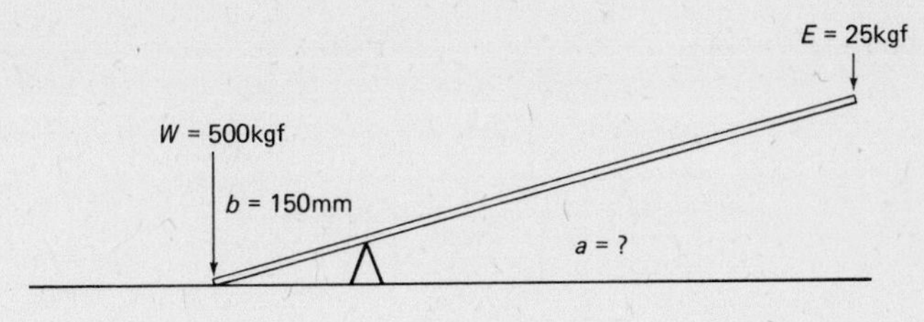

Fig. 18

Formula = $Wb = Ea$.
Rearranging formula to suit:

$$a = \frac{Wb}{E}$$

Substituting letters for numbers:

$$a = \frac{500 \times 0.15}{25}$$

$$a = \frac{75}{25}$$

Answer $a = 3$ m

Another application of the use of levers in plumbing is when we use the bending machine to form copper or steel tubing.

The centre pin of the machine acts as the fulcrum. The backslide as the point load, and the effort is the handle where we apply the pull.

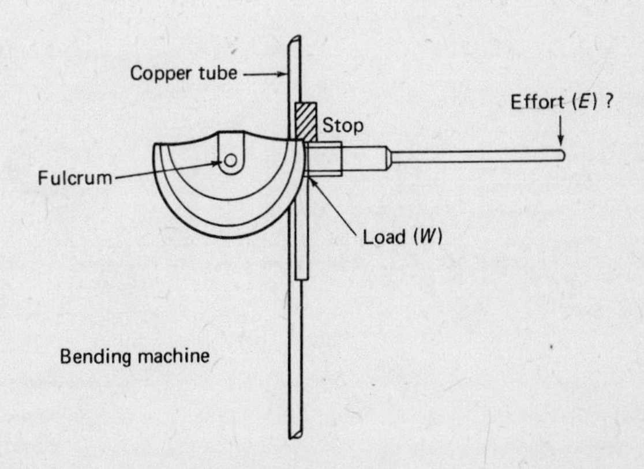

Fig. 19

This can be simplified to:

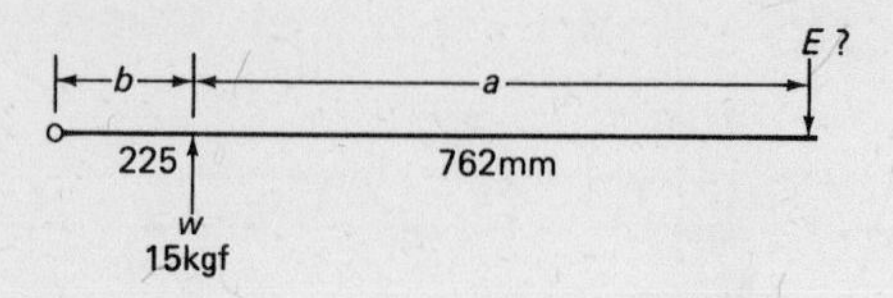

Fig. 20

To find effort (E).
Formula = $Wb = Ea$
Rearranging the formula to suit:

$$E = \frac{Wb}{a}$$

Substituting letters for numbers:

$$E = \frac{15 \times 0.225}{0.762}$$

$$E = \frac{3.375}{0.762}$$

Answer $E = 4.429$ kgf

In some plumbing materials, the principles of levers is also used. A good example is the ballvalve. The sketch shows the relevant point with assumed lengths and loads:

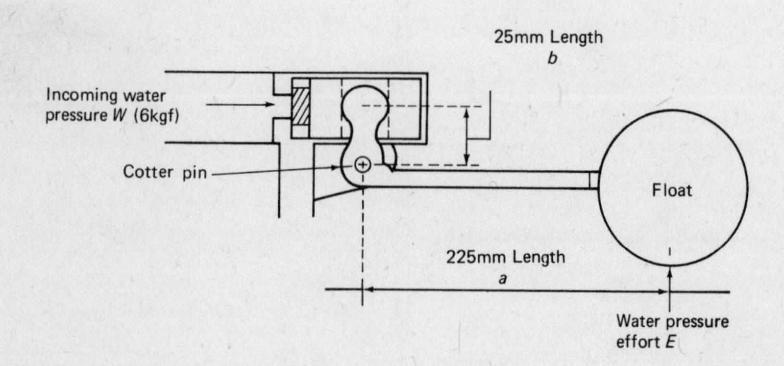

Fig. 21

This can be simplified to Figure 22.

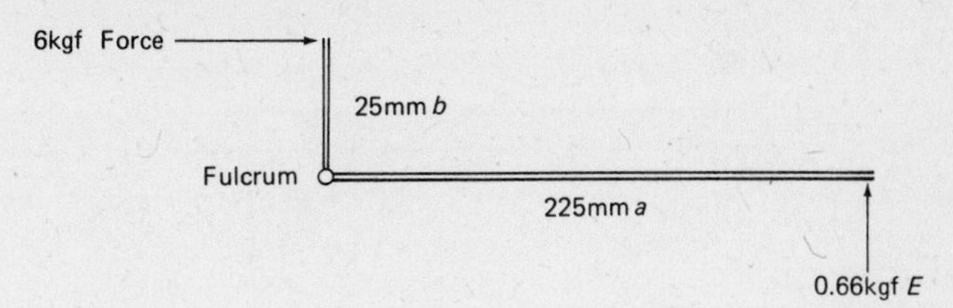

Fig. 22

To find effort (E).
Formula $Wb = Ea$.
Rearranging the formula to suit:

$$E = \frac{Wb}{a}$$

Substituting numbers for letters:

$$E = \frac{6 \times 0.025}{0.225}$$

$$E = \frac{0.150}{0.225}$$

Answer $E = 0.66$ kgf

REMEMBER to convert kgf to newtons multiply by 9.81.
i.e. $E = 0.66 \text{ kgf} = 0.66 \times 9.81$
$E = 6.4746$ N.

To find length a.
Formula = $Wb = Ea$.
Rearranging the formula to suit:

$$\frac{Wb}{E} = a$$

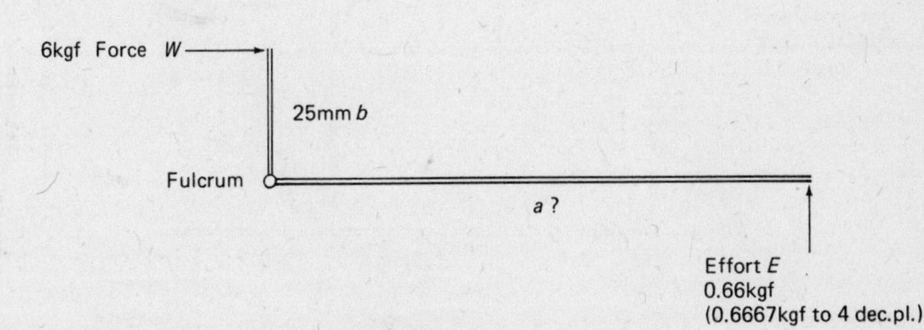

Fig. 23

Substituting numbers for letters:

$$a = \frac{6 \times 0.025}{0.66}$$

$$a = \frac{0.150}{0.6667}$$

Answer $a = \underline{225 \text{ mm}}$

To find length b.
Formula = $Wb = Ea$.
Rearranging the formula to suit:

$$\frac{E \times a}{W} = b$$

Substituting numbers for letters:

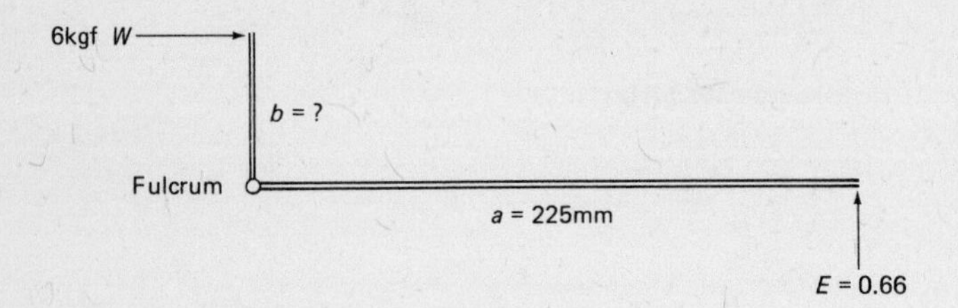

Fig. 24

$$b = \frac{0.6667 \times 0.225}{6}$$

$$b = \frac{0.150}{6}$$

$$b = 0.025$$

Answer $b = 25 \text{ mm}$

To find load W.

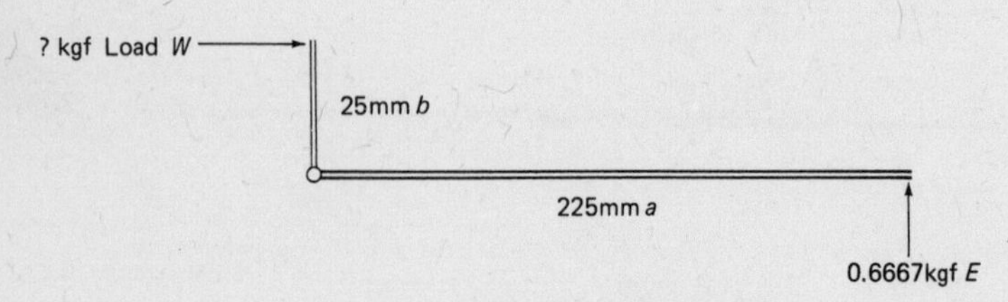

Fig. 25

Rearranging formula to suit:

$$W = \frac{E \times a}{b}$$

Substituting numbers for letters:

$$W = \frac{6667 \times 0.225}{0.025}$$

$$W = \frac{0.150}{0.025}$$

Answer $W = 6$ kgf

So far the examples looked at have involved the action of a Portsmouth type Ballvalve, now we will look at the action found with the Croydon type Ballvalve which is a little different.

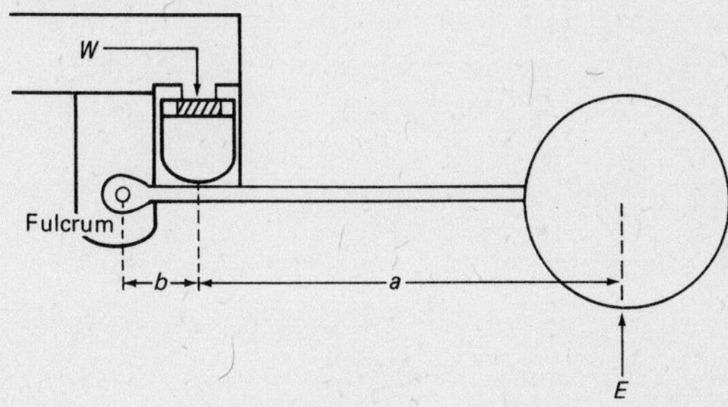

Fig. 26

To find load W.
Formula = $Wb = Ea$.
Rearranging the formula to suit:

$$W = \frac{E \times a}{b}$$

$$W = \frac{0.6667 \times 0.225}{0.025}$$

$$W = \frac{0.150}{0.025}$$

Answer $W = 6$

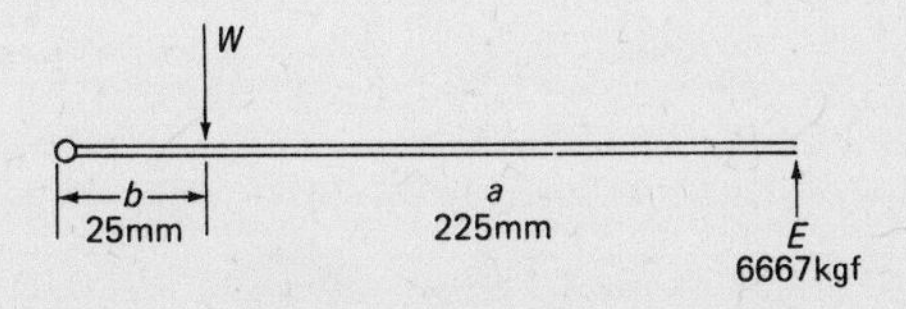

Revision questions

1. A large crate is to be lifted at one end by a lever. The force created by the crate is 700 kgf. The distance from the crate to the fulcrum is 200 mm and the distance from the fulcrum to the end of the lever is 900 mm long. Determine the force required at the free end of the lever to raise the crate.
2. A Portsmouth pattern ballvalve has a lever arm 250 mm long, and the distance between the centre of the valve and the cotter pin is 30 mm. The water pressure on the float is 80 kgf. Determine the force exerted by the water on the seating of the valve.

14 Water pressure

Data

1 cm^3 of water at 4°C has a mass of 1 gramme (g)
1 m^3 of water contains 100 x 100 x 100 cm^3 (1×10^6 cm^3)
1 m^3 of water contains 1,000,000 dm^3 (10^6 cm^3)
1 m^3 of water contains 10^6 cm^3 and therefore 10^6 grammes in mass
1 m^3 of water has a mass of 1000 kilogrammes (kg)

To find the force created by a given mass of water the formula $\frac{\text{kg}}{\text{area}}$ is used.

1 kilogramme force is the force produced by 1 kg mass of water

Alternatively the force may be expressed in newtons, 1 kgf = 9.81 newtons.

It follows then that if the kgf is known, it may be converted into newtons by multiplying by 9.81.

Above we see that 1 m^3 of water = 1000 kg, and that 1 kg of water will produce 1 kgf, therefore 1000 kg will produce 1000 kgf. If this mass is then multiplied by 9.81 it will give newtons.

1000 kgf x 9.81 = 9810.0 newtons.

Newtons like grammes are rather small giving rise to the use of kilogrammes or kilonewtons.

To convert grammes or newtons to kilogrammes or kilonewtons divide by 1000.

$$\frac{9810.0 \text{ N}}{1000} = 9.81 \text{ kN}$$

$$\frac{9810.0 \text{ g}}{1000} = 9.81 \text{ kg}$$

Figure 27 shows a 1 dm^2 area column having a force of 98 newtons at its base, and a height of 1 m. It can be seen that an area of 1 dm^2 will divide into the area of 1 m^2 exactly 100 times. The force acting on a 1 m^2 area will then be 100 x 98 = 9800 newtons (9.8 kN).

If the height of the column is increased to 2 m, the pressure is doubled at the base.

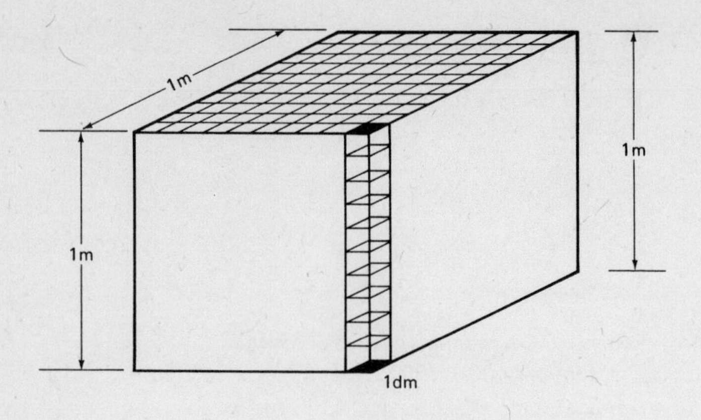

Fig. 27

Pressure of water at the base of 1 m^2 area created by a column of water 2 m high

$= 2 \times 9.81$ kN

$= 19.62$ kN

Note 1 dm^3 equals 1 litre,
1 dm^3 equals 1 kg (at 4°C)

Force

The name for force is the newton, which is the surname of the scientist who discovered it, and derives from

$$\frac{\text{kg m}}{\text{s}^2} = \text{newton}.$$

The gravitational pull of the earth creates an acceleration on a falling body of 9.81 m/s^2, and therefore the force in newtons on a mass of 1 kg = 9.81 m/s^2, giving us

1 kgf = 9.81 newtons

Worked examples

1. Calculate the force in newtons acting on the base of a cold water storage cistern measuring 400 mm × 400 mm × 400 mm.

Formula mass x 9.81
= 0.4 m x 0.4 m x 0.4 m (m^3)
= 0.4 m x 0.4 m x 0.4 m x 1000 (kg)
= 0.4 m x 0.4 m x 0.4 m x 1000 x 9.81 (newtons)
then 0.16 x 0.4 x 1000 x 9.81
= 0.064 x 1000 x 9.81
= 64 x 9.81
= 627.84 newtons
or 0.628 kilonewtons.

Note 1 m^3 of water has a mass of 1000 kg
1 kgf = 9.81 newtons.

2. Calculate the force in newtons exerted on the base of a cold water storage cistern containing 400 kg of water.
Formula mass x 9.81
= 400 kg x 9.81
= 3924 newtons
or 3.924 kilonewtons.

Intensity of pressure

This is the measure of a force on a given area.

Worked examples

1. Calculate the intensity of pressure on the base of a cold water storage cistern measuring 450 mm x 450 mm x 600 mm high.

Formula $\frac{\text{mass} \times 9.8}{\text{area}}$ (for ease of calculation the digit 1 has been omitted)

$$= \frac{0.45 \text{ m} \times 0.45 \text{ m} \times 0.6 \text{ m} \times 1000 \times 9.8}{0.45 \text{ m} \times 0.45 \text{ m}}$$

Note 1000 kg = 1 m^3 of water.

Cancelling out.

$$= \frac{\cancel{0.45} \text{ m} \times \cancel{0.45} \text{ m} \times 0.6 \text{ m} \times 1000 \times 9.8}{\cancel{0.45} \text{ m} \times \cancel{0.45} \text{ m}}$$

= 0.6 m x 1000 x 9.8
= 5.88 x 1000
= 5880 N/m^2
or 5.88 kN/m^2

Note Formula for N/m^2 = head (m) x 1000 x 9.8.

$$\text{Formula for } kN/m^2 = \frac{\text{head (m)} \times 1000 \times 9.8}{1000}$$

Therefore if both 1000 are cancelled out we are left with head (m) x 9.8 = kN/m^2.

2. Calculate the intensity of pressure on the base of a hot water storage cylinder measuring 500 mm in diameter, and 1 m in height.
   Formula head (m) x 9.8
   = 1 m x 9.8
   = 9.8 kN/m^2

In practice however we do not always deal with areas of 1 m^2, one such case is that of a seating of a tap.

Worked examples

1. Calculate the force exerted on the seating of a 25 mm tap fed from a head of 5 metres (newtons).
   Formula mass x 9.8
   = area of tap x head x 1000 x 9.8
   = πr^2 x head (m) x 1000 x 9.8
   = 3.142 x 0.012 x 0.012 x 5 x 1000 x 9.8
   = 0.000452 x 5 x 1000 x 9.8
   = 0.452 x 5 x 9.8
   = 2.262 x 9.8
   = 22.170 newtons

2. Calculate the intensity of pressure exerted on a 25 mm seating of a tap ($newtons/m^2$).
   Formula head (m) x 1000 x 9.8
   = 5 m x 1000 x 9.8
   = 49 x 1000
   = 49 000 $newtons/m^2$

3. Calculate the force exerted on the seating of a 25 mm tap fed from a head of 3 metres (kN)
   Formula head (m) x area (m) x 9.8
   = 5 m x πr^2 x 9.8
   = 5 m x 3.142 x 0.012 x 0.012 x 9.8
   = 0.000453 x 5 x 9.8
   = 0.022169 kN

When dealing with PRESSURE we may be working in millibars for very accurate work, but in most cases we will work in newtons per square metre, or in newtons per metre square, kilonewtons per metre square, bars or even kilograms per cubic metre. Table 5 enables the conversion of newtons per metre square to millibars, kilonewtons to bars, and table 6 gives the density of water at several different temperatures.

Table 5

Pressure

Newtons/m^2	millibar
100	1
1000	10
1 kN/m^2	10
10 kN/m^2	100
20 kN/m^2	200
30 kN/m^2	300
40 kN/m^2	400
50 kN/m^2	500
55 kN/m^2	550
100 kN/m^2	1000
100 kN/m^2	1 bar
200 kN/m^2	2 bar
300 kN/m^2	3 bar
400 kN/m^2	4 bar
500 kN/m^2	5 bar
550 kN/m^2	5.5 bar
1000 kN/m^2	10 bar
1 MN/m^2	10 bar

Table 6

Density of water

Celsius	kg/m^3
0	999.80
4	1000.00 maximum density
10	999.70
15.6	990.00
21.1	997.99
26.7	996.63
32.2	994.98
37.8	993.07
43.3	990.93
48.9	988.56
54.4	986.05
60.00	983.24
65.6	980.29
71.1	977.13
76.7	973.88
82.2	970.43
87.8	966.82
93.3	963.07
100.0	958.38 boiling point.

Conversion factor, 1 kN/m^2 = 0.102 metres head.

Worked examples

An example of the use of millibars is when we use a manometer to test building services.

1. A manometer reading is 50 mm, convert this reading to millibars.
   If 1 m head = 9.81 kN/m^2
   Then 50 mm head will = 9.81 x 0.05 kN/m^2
   Therefore 50 mm head = 0.49 kN/m^2

From the tables, 1 kN/m^2 = 10 mb

So 0.49 x 10 = 4.9 mb

We may of course be given a problem where the head in metres is not given, but must be found from a given pressure.

2. The reading on a pressure gauge is 50 kN/m^2, convert this reading to metres.

We know that 9.81 kN/m^2 = 1 metre in height

Therefore

$$\frac{50\ kN/m^2}{9.81\ kN/m^2} = 5.1 \text{ metres head}$$

3. Alternatively, using the conversion factor above

1 kN/m^2 = 0.102 metres head

Therefore

50 kN/m^2 x 0.102 = 5.1 kN/m^2

Table 7 *Conversion of metres head to kN/m² and mb (bars)*

Head in metres	*kN/m²*	*mb and bars*
1	9.81	98 mb
2	19.62	196 mb
3	29.43	294 mb
4	39.24	392 mb
5	49.05	490 mb
6	58.86	588 mb
7	68.67	686 mb
8	78.48	784 mb
9	88.29	883 mb
10	98.10	981 mb
20	196.20	1.96 bar
30	294.30	2.94 bar
40	392.40	3.92 bar
50	490.50	4.90 bar
100	981.00	9.81 bar

Revision questions

1. Calculate the force exerted by water on the base of a cold water storage cistern measuring 2 m x 2 m x 3 m high.
2. Calculate the intensity of pressure on the base of a hot water storage cylinder measuring 500 mm in diameter and 1.5 m high.

15 Box's formula

This is a formula used by plumbers to calculate one of the following in a set condition.

Diameter of pipe	(in centimetres)
Length of pipe	(in metres)
Flow rate	(in litres per second)
Head	(in metres)

Box's formula is generally expressed as

$$q = \frac{1}{5}\sqrt{\frac{d^5 h}{l}}$$

and has an allowance for frictional loss included.

The allowed frictional loss is acceptable for small domestic pipes of smooth bore only, and is rated as 0.0075.

Obviously to calculate all four requirements the formula will have to be rearranged to suit as follows.

$$q = \frac{1}{5}\sqrt{\frac{d^5 h}{l}} \text{ to } h.$$

$$(5q)^2 = \frac{d^5 h}{l}$$

$$\frac{l(5q)^2}{d^5} = h$$

$$q = \frac{1}{5}\sqrt{\frac{d^5 h}{l}} \text{ to } l$$

$$(5q)^2 = \frac{d^5 h}{l}$$

$$l(5q)^2 = d^5 h$$

$$l = \frac{d^2 h}{(5q)^2}$$

$$q = \frac{1}{5}\sqrt{\frac{d^5 h}{l}} \text{ to } d$$

$$(5q)^2 = \frac{d^5 h}{l}$$

$$\frac{l(5q)^2}{h} = d^5$$

$$\sqrt[5]{\frac{l(5q)^2}{h}} = d$$

Examples

1. A cold water storage cistern is situated 4 metres above a 12 mm bib tap, and has an 8 metres run of pipe. Calculate the discharge in litres/second expected from this tap.

Formula

$$q = \frac{1}{5}\sqrt{\frac{d^5 h}{l}}$$

$$q = \frac{1}{5}\sqrt{\frac{1.2^5 4}{8}}$$

$$q = \frac{1}{5}\sqrt{1.24415}$$

$q = 0.2231$ litres/second
(electronic calculator)

No	Log.
1.2^5	0.0792 x 5
	0.3960 +
4	0.6021
	0.9981 −
8	0.9031
$\sqrt[5]{}$	0.0950 ÷ 2
	0.0475 −
	0.6990
Antilog.	$\bar{1}.3485$
Answer	0.2231

(logarithms)

2. If the discharge is 0.22308 litres/sec from a 12 mm bib tap, with a head of 4 metres, what is the length of pipe involved?

Formula

$$L = \frac{d^5 h}{(5q)^2}$$

$$L = \frac{1.2^5 4}{(5 \times 0.22308)^2}$$

$$L = \frac{9.9532}{1.2441}$$

$L = 8$ metres

No	Log.	
1.2^5	0.0792 x 5	
	0.3960 +	
4	0.6021	
	0.9981	0.9981
$(5 \times 0.2231)^2$	0.6990 +	−
	$\bar{1}.3485$	
	0.0475 x 2	
	0.0950	0.0950
		0.9031
Antilog.	0.9031	
Answer	8.0000	

3. What diameter of pipe would be required to give a discharge of 0.22308 litres/sec through a pipe 8 metres long, from a cold water cistern positioned 4 metres above the bib tap.

Formula

$$d = \sqrt[5]{\frac{(5 \times 0.22308)^2 \times 8}{4}}$$

$$d = \sqrt[5]{\frac{1.2441 \times 8}{4}}$$

$$d = \sqrt[5]{2.4882}$$

Answer $d = 1.1999$ centimetres.

No	Log.
$(5 \times 0.22308)^2$	0.6990 +
	$\bar{1}.3485$
	0.0475 x 2
	0.0950 +
8	0.9031
	0.9981 −
4	0.6021
$\sqrt[5]{}$	0.3960 ÷ 5
Antilog.	0.0792
Answer	1.200

4. If a 12 mm diameter pipe to a 12 mm bib tap is expected to deliver 0.22308 litres/sec when the length of the pipe is 8 metres, what head is required?

Formula

$$\frac{L \times (5q)^2}{d^5} = h$$

$$h = \frac{8 \times (5 \times 0.22308)^2}{1.2^5}$$

$$h = \frac{9.9538}{2.4883}$$

$h = 3.9998$ metres

Answer 4 metres

No	Log.	
$(5 \times 0.22308)^2$	0.6990 +	
	$\bar{1}.3485$	
	0.0475 x 2	
	0.0950 +	
8	0.9031	
	0.9981	0.9981
$(1.2)^5$	0.0792 x 5	−
	0.3960	0.3960
		0.6021
Antilog.	0.6021	
Answer	4.000	

Revision questions

1. A 22 mm diameter pipe is fed from a cistern situated 4 m above it. The length of the pipe feeding the tap is 5.5 m long. Calculate the discharge in litres per second from the tap.
2. If the discharge from a 22 mm diameter pipe is 1.2247 litres/second fed from a height of 5 m find the length of the pipe involved.

16 Chezy formula

This formula is used in drainage and is expressed as $V = C\sqrt{mi}$.
Where:

V = Velocity in metres/second
C = Co-efficient
m = Hydraulic mean depths
i = Incline or fall

As with any formula we will need to be able to rearrange it to suit our needs, so that knowing three of the quantities we can find the fourth. Rearranging the formula from $V = C\sqrt{mi}$ to suit.

Rearranging $V = C\sqrt{mi}$ to find i

$$\frac{V}{C} = \sqrt{mi}$$

$$\frac{V^2}{C^2} = mi$$

$$\frac{V^2}{C^2 m} = i$$

Rearranging $V = C\sqrt{mi}$ to find C

$$\frac{V}{\sqrt{mi}} = C$$

Rearranging $V = C\sqrt{mi}$ to find m

$$\frac{V^2}{C^2} = mi$$

$$\frac{V^2}{C^2 i} = m$$

Rearranging $\frac{V^2}{C^2 m} = i$ to find V

$$\frac{V^2}{C^2} = mi$$

$$\frac{V}{C} = \sqrt{mi}$$

$$V = C\sqrt{mi}$$

Now we can rearrange the formula to suit our needs we must also consider the letter m in the formula. The value of m may or may not be given.

m is the relationship between amount of liquid flowing in a pipe and its contact with the internal bore of that pipe. The formula for finding m is:

$$\frac{\text{Cross sectional area of flow (mm}^2\text{)}}{\text{Length of wetted perimeter (mm)}}$$

Worked example

If the pipe is flowing full, assume the pipe diameter to be 100 mm.

Solution:

$$m = \frac{\text{Cross sectional area of flow (mm}^2\text{)}}{\text{Length of wetted perimeter (mm)}}$$

$$m = \frac{\pi r^2}{\pi d}$$

or may be expressed as

$$\frac{\frac{\pi d^2}{4}}{\pi d}$$

Therefore: $\dfrac{\pi r^2}{\pi d}$

Cancel out: $\dfrac{\not{\pi} r^2}{\not{\pi} d}$

$\dfrac{r^2}{d}$

Substituting: $= \dfrac{50 \times 50}{100}$

$= \dfrac{50}{2}$

Answer m or HMD = 25

or

Therefore: $\dfrac{\pi d^2}{4} \div \pi d$

or $\dfrac{\pi d^2}{4} \times \dfrac{1}{\pi d}$

Cancel out: $= \dfrac{\not{\pi} d^2}{4} \times \dfrac{1}{\not{\pi}\not{d}}$

$= \dfrac{d}{4}$

Substituting: $= \dfrac{100}{4}$

Answer m or HMD = 25

If the drain is running half full and of the same diameter then the answer would be the same. This is because the cross sectional area of flow and the length of wetted perimeter are both halved.

2. A 100 mm diameter pipe is running half full.

Formula: $\dfrac{\text{Cross sectional area of flow (mm}^2\text{)}}{\text{Length of wetted perimeter (mm)}}$

$$\frac{\frac{\pi r^2}{2}}{\frac{\pi d}{2}} \quad \textit{or} \quad \frac{\pi r^2}{2} \div \frac{\pi d}{2}$$

$$= \frac{\pi r^2}{2} \times \frac{2}{\pi d}$$

Cancel out: $\dfrac{\not{\pi} r^2}{\not{2}} \times \dfrac{\not{2}}{\not{\pi} d}$

$$= \frac{r^2}{d}$$

Substituting: $\dfrac{r^2}{d} = \dfrac{50 \times 50}{100} = \dfrac{50}{2}$

Answer 25

Obviously a drain could be running at various depths, and to find the HMD of such a drain would need much more complicated calculations than that which we have just practised, but these cannot be dealt with within the scope of this book.

Now we can find the HMD we can consider working out the whole formula of Chezy. The co-efficient C is generally given from tables to suit the diameter of pipe and also its material, incline and age.

Worked example

1. Calculate the velocity V of flow in a new C.I. drain flowing full bore and laid at a fall of 1 in 60. The diameter of the drain is 150 mm and the co-efficient C is taken as 1.9

Formula: $V = C\sqrt{mi}$

Note. HMD or $m = \dfrac{\not{\pi} r^2}{\not{\pi} d} = \dfrac{r^2}{d} = \dfrac{75 \times 75}{150} = 37.5$

Substituting

$$V = 1.9\sqrt{37.5 \times 1/60}$$
$$V = 1.9\sqrt{37.5 \div 60}$$
$$V = 1.9\sqrt{0.625}$$
$$V = 1.9 \times 0.79$$

Answer $V = 1.5$ metres per second

2. Find the co-efficient C of a drain running full bore at a fall of 1/60. The pipe is 150 mm in diameter and discharges 1.5 metres per second, with an HMD of 37.5.

Formula: $V = C\sqrt{mi}$ rearrranged to $C = \dfrac{V}{\sqrt{mi}}$

Substituting

$$C = \frac{1.5}{\sqrt{37.5 \times 0.1066}}$$

$$C = \frac{1.5}{0.791}$$

$$C = 1.896$$

Answer $C =$ say 1.9

Revision questions

1. A drain is layed at a fall of 1:40 and is 100 mm in diameter. If the co-efficient of the drain is 1.9, calculate the velocity of flow. The HMD is 37.5.
2. Find the co-efficient of a 100 mm diameter drain laid at a fall of 1 in 40. The HMD is 37.5.

17 Logarithms

Logarithms (logs), are catalogued digits, used in mathematics to speed up the calculations without increasing errors. To use logs, the actual digits are changed for those in the log tables, then on completion of the calculation are converted back to actual figures.

Before looking for the log of a number, a characteristic is found first. The characteristic determines the position of the decimal point in the final answer.

The characteristic of an actual number is found by deducting one digit less than the number of digits to the left of the decimal point, in the actual number, i.e.

The characteristic of 7654. = 3. (7654 = 4 digits, less one = 3)
The characteristics of 765. = 2. (765 = 3 digits, less one = 2)
of 76. = 1. (76 = 2 digits, less one = 1)
of 7. = 0. (7 = 1 digit, less one = 0)
of 0. = $\bar{1}$. (0 = no digits, less one = –1)
of 0.0 = $\bar{2}$. (0.0 = no digits to the left of the decimal point, and the first digit of the decimal is also missing = –2)
of 0.00 = $\bar{3}$. and so on.

Now the characteristic can be found, the log of the number may be found from the log tables.

At the head of the log tables there is a row of digits in large heavy print, numbering from 0 to 9, and then continued in smaller digits reading from 1 to 9. These smaller digits represent the *mean difference* column. Down the left hand side of the page is a column of digits reading from 10 to 99, generally over two pages.

To find the log of a number such as 765 (first find the characteristic, 2), look down the left hand side of the page for the number 76, this being the first two digits of the actual number. Now look along the top row for the figure 5, where the two rows meet will be the answer, 8837. (This number is called the *mantissa*.)

When using logs the following system is generally used.

Worked examples

1. Find the log of 765.

No.	Log
765	2.8837

The number 765 goes under the No. column, and the characteristic and the log are placed under the log column.

2. Find the log of 76.

No.	Log
76	1.8808

Because there is no third digit the log of 76 is found under the column headed 0.

3. Find the log of 7.

No.	Log
7	0.8451

It can be seen that there is no digit 7 in the left hand column. The log for 7 being the same as that for 70 or 700 or 7000 etc. The characteristic makes the adjustment.

4. Find the log of 0.7.

No.	Log
0.7	$\bar{1}.8451$

The log remains the same as for the digit 7, but the characteristic is different.

5. Find the log of 10.

No.	Log
10	1.0000

The log of 10 is 0000, but the characteristic is 1. The characteristic will increase the final answer by one decimal place, this is the same as multiplying 6 x 10 to equal 60. The decimal point only has been moved one place to the right. The same will apply to the digits 100, 1000, etc., a characteristic will be shown only.

All the examples shown use digits up to three decimal places, 765. To find the log of a digit greater than three decimal places, the mean difference column is used along with the digits in large print already used.

6. To find the log of a number such as 8765.
   1. First find the characteristic, 3.
   2. Look for the number 87, down the left hand column.
   3. Find the digit 6 on the top row in large print.

4. Where these two numbers meet, the number will read .9425. (Write this number down.)
5. Still on the row beginning with the digits 87, move under the mean difference column, under the number 5.
6. Where these two numbers meet, is found the digit 2.
7. Add this digit to the first number found, 9425, and this will be the log of 8765 = 9427.

7. The log of 5432 will be,

No.	Log
5432	3.7350

The log of 543 = 7348
plus 2
7350

When the log of a number such as 65432 is required, the last digit is ignored, this is because the log tables are printed to four decimal places only. If the digit to be ignored is greater than the digit 5, then the preceeding digit is increased by one, 65476 would be 6548.

The log of a number which has decimals in it, is found by ignoring the decimal point, once the characteristic has been found.

8. The log of 654.8 would be,

No.	Log
654.8	2.8161

The characteristic of 654.8 = 2
The log of 654 = .8156
plus 5
.8161

9. The log of 65.48 would be,

No.	Log
65.48	1.8161

The number does not alter, only the characteristic.

10. The log of 2.61 would be,

No.	Log
2.61	0.4166

11. The log of 0.261 would be,

No.	Log
0.261	$\bar{1}.4166$

12. The log of 0.0261 would be,

No.	Log
0.0261	$\bar{2}.4166$

The *antilogarithm* of a number is found exactly in the same way as for finding the logarithm, only this time we are converting logs back to actual numbers.

To find the antilogarithm (antilog), of 2.6100, turn the page of log tables over to the next page headed antilogs.

On the top of the page there are large digits in heavy print, reading from 0 to 9 and then continuing in smaller digits reading from 1 to 9 in the mean difference column. Down the left hand side of the page is a column of numbers reading from .00 to .99 on the next page.

To antilog the number 2.6100, ignore the characteristic of 2, and find the antilog in exactly the same way as finding the log.

The antilog of .6100 = 4074.

What we do not know is the value of the number 4074, it may be 4.074, 40.74, 407.4 or a whole number of 4074.

To determine the true value of the number we consult the characteristic that we ignored earlier, 2. The characteristic tells us where to insert the decimal point in our answer. The following chart will have to be remembered.

Characteristic		*Position in answer*
$\bar{3}$.	=	0.00123
$\bar{2}$.	=	0.0123
$\bar{1}$.	=	0.123
0.	=	1.23
1.	=	12.3
2.	=	123.0
3.	=	1230.0 and so on.

13. Find the log of 80 and then antilog same.

No.	Log
80	1.9031

Antilog 8000
Answer 80.0
(Antilog of 903 = 7998 + 2 = 8000)
(Characteristic of 1 = two decimal places)

14. Find the antilog of 4.5432.

No.	Log
	4.5432
Antilog	3493
Answer	34930.0

(Antilog of 543 = 3491 + 2 = 3493.
Characteristic of 4 = five decimal places.
Because there were only four digits, and five decimal places a 0 is added to complete the number.)

15. Find the log of 0.5432 and antilog same.

No.	Log
0.5432	$\bar{1}.7350$
Antilog	5433
Answer	0.5433

(Notice there is a mathematical error of 0.0001 which is acceptable.)

There are four main uses for logarithms:

multiplication
division
finding powers of numbers
finding roots of numbers

In logs to add is to multiply and to subtract is to divide.

Worked examples

Multiplication

16. Multiply 12 x 12.

No.	Log
12	1.0792 +
12	1.0792
Antilog	2.1584
Answer	1440
	144.0

(Antilog of 158 = 1439 + 1 = 1440)
(Characteristic of 2 = three decimal places)

17. Multiply 432 x 234.

No.	Log	
432	2.6355	+
234	2.3692	
	5.0047	
Antilog	1011	
Answer	101100.0	

(Antilog of .0047 is found by looking down the left hand side of the tables for .00, first line.)
(Characteristic of 5 = six decimal places)

18. Multiply 0.123 x 00.0012.

No.	Log	
0.123	$\bar{1}.0899$	+
0.0012	$\bar{3}.0792$	
	$\bar{4}.1691$	
Antilog	1476	
Answer	0.0001476	

(Characteristic of $\bar{4}$ = three decimal places in front of the number, this means adding three noughts in front of the number.)

Division

19. Divide 432 by 6.

No.	Log	
432	2.6355	–
6	0.7782	
	1.8573	
Antilog	7199	
Answer	71.99 say 72.0	

(Antilog of 857 = 7194 + 5 = 7199)

20. Divide 65 by 120.

No.	Log	
65	1.8129	–
120	2.0792	
	$\bar{1}.7337$	(2 from 1 = –1)
Antilog	5417	
Answer	0.5417	

21. Divide 0.543 by 0.032.

No.	Log
0.543	$\bar{1}.7348$ −
0.032	$\bar{2}.5051$
	$\bar{0}.2297$ ($\bar{1} - \bar{2} = 1$.)
Antilog	1697
Answer	16.97

22. Divide 400 by 20.

No.	Log
400	2.6021 −
20	1.3010
	1.3011
Antilog	2000
Answer	20.0

Powers

To find the value of 8^2, find the log of 8, multiply by 2 and antilog the answer.

23. Find the value of 8^2.

No.	Log
8^2	0.9031 x 2
=	1.8062
Antilog	6400
Answer	64.0

24. Find the value of 12^2.

No.	Log
12^2	1.0792 x 2
	2.1584
Antilog	1440
Answer	144.0

25. Find the value of 6^3, find the log, multiply by 3, and antilog the answer.

No.	Log
6^3	0.7782 x 3
	2.3346
Antilog	2161
Answer	216.1

26. Find the value of 16^5.

No.	Log
16^5	1.2041×5
	6.0205
Antilog	1048
Answer	1048000.0

27. Find the value of 0.056^2.

No.	Log	
0.056^2	$\bar{2}.7482 \times 2$	
	$\bar{4}.4964$	$(2 \times -2 = -4 \text{ less } 1 = -3)$
Antilog	3136	
Answer	0.003136	

Roots

To find the square root of a number we use the reverse system that we used for finding the power of a number.

To find the square root, or any other root, we adopt the same procedure, find the log of the number, divide by the root and antilog the answer.

28. Find the square root of 64.

No.	Log
$\sqrt{64}$	$1.8062 \div 2$
	0.9031
Antilog	8000
Answer	8.0

29. Find the cube root of 216.

No.	Log
$\sqrt[3]{216}$	$2.3345 \div 3$
	0.7781
Antilog	5999
Answer	6.0

30. Find the square root of 0.0876.

No.	Log
$\sqrt{0.0876}$	$\bar{2}.9425 \div 2$
	$\bar{1}.4712$
Antilog	2959
Answer	0.2959

31. Find the number of litres that could be contained in a hot water storage cylinder measuring 1 metre in diameter and 3 metres in height.

Formula:

$$\pi r^2 h \times 1000 = \text{litres}$$

Using logs:

No.	Log
(π) 3.142	0.4972
(r) 0.5	$\bar{1}.6990$
(r) 0.5	$\bar{1}.6990$
(h) 3.0	0.4771
1000	3.0000
Antilog	3.3723
	2357
Answer	2357 litres

32. Find the radius when the height of the hot water cylinder is 3 metres, and the number of litres contained in it is 2357.

Formula:

$$r = \sqrt{\frac{\text{litres}}{\pi \times h \times 1000}}$$

$$r = \sqrt{\frac{2357}{3.142 \times 3 \times 1000}}$$

Using logs:

	No.	Log	
(litres)	2357	3.3724	3.3724
(π)	3.142	0.4972	
(h)	3.0	0.4771 +	
	1000	3.0000	
		3.9743	3.9743
			$\bar{1}.3981$
	$\sqrt{}$ =	$\bar{1}.3981 \div 2$	
	Antilog	$\bar{1}.6990$	
	Answer	0.5 m = radii	

N.B. When multiplying by 1000 to convert cubic capacity to litres, the cubic capacity must be in *metres cubed.*

33. Find the height of the hot water cylinder when the diameter of the cylinder is 1 metre, and the number of litres held in it, is 2357.

Formula:

$$h = \frac{\text{litres}}{\pi r^2 \times 1000}$$

$$h = \frac{2357}{3.142 \times 0.5 \times 0.5 \times 1000}$$

Using logs:

	No.	Log	
(litres)	2357	3.3724	3.3724
(π)	3.142	0.4972	
(r)	0.5	$\bar{1}.6990$ +	
(r)	0.5	$\bar{1}.6990$	
	1000	3.0000	
		2.8952	2.8952
			0.4772
	Antilog	0.4772	
	Answer	2.999 metres	
	say	3 metres	

Revision questions

Find the characteristics of:

1. 94 2. 0.66 3. 1068

Find the logarithm of:

4. 18 5. 4 6. 0.81

Find the anti-logarithm of:

7. 8163 8. 1438 9. 2000 10. $\bar{1}.3843$

Multiply the following:

11. 78 x 14 12. 21.6 x 18.36

Divide the following:

13. 17 ÷ 4 14. 216 ÷ 61

Find the square root of:

15. 87.6 16. 14.31 17. 91

18 Temperature scales and conversion tables

Fahrenheit scale	*Reaction*	*Celsius scale*	*Comment*
212	Boiling point	100.0	Rapid expansion in volume 1700 times that of water (approx.)
200		93.3	
185	Primary flow pipe temp. h.w.	85.0	Domestic dwellings. Hot water
180		82.2	
170		76.7	
167	Radiator mean temp.	75.0	Domestic or public
160		71.1	
149	Primary return pipe temp.	65.0	Domestic dwellings. Hot water
140	Washing up water temp.	60.0	Domestic dwellings
120		48.9	
110	Bath water temperature	43.3	Domestic dwellings and public
105		40.6	
104		40.0	
103		39.4	
102		38.9	
100		37.8	
99		37.2	
98		36.7	
97		36.1	
96		35.6	
95		35.0	
90		32.2	
80		26.7	
70		21.1	
68	Av. room temperature	20.0	Domestic dwellings
60		15.6	
50		10.0	
40	Maximum density of pure	4.4	
	water at sea level	4.0	Pure water at sea level
32	Freezing point of water	0.0	Pure water at sea level Rapid expansion of water due to freezing, about 1/10

N.B. 4°C is the maximum density of pure water at ordinary temperature at sea level, any rise or fall in temperature or impurity addition to the water will result in an alteration in reading.

The maximum density of water is the temperature at which water occupies its least volume. Any rise or fall in temperature will result in an expansion of the water.

The name *celsius* is the new name given to the existing *centigrade* scale, and is in fact the surname of the man who invented the centigrade scale. The Celsius scale is graduated in degrees from 0 to 100°.

Conversion factor

To convert Fahrenheit to Celsius, multiply by $\frac{5}{9}$ after deducting 32°.

Example Convert 70°F to °C.

$(70 - 32) \times \frac{5}{9}$

$38 \times \frac{5}{9}$

$\frac{190}{9}$

Answer 21.1°C

To convert Celsius to Fahrenheit, multiply by $\frac{9}{5}$ then add 32°. (The addition of 32 may be left until last if preferred.)

10°C to °F

$10 \times \frac{9}{5} + 32$

$\frac{90}{5} + 32$

$18 + 32$

Answer 50°F

19 Conversion tables

1. Inches to millimetres linear conversion.

Multiply by 25.4. 1 metre = 39.370 ins.

Ins/mm	*Ins/mm*	*Ins/mm*	*Ins/mm*
$\frac{1}{64}$: 0.3969	3 : 76.1999	15 : 381.0000	27 : 685.7999
$\frac{1}{32}$: 0.7838	4 : 101.6000	16 : 406.4000	28 : 711.1999
$\frac{1}{16}$: 1.5875	5 : 127.0000	17 : 431.8000	29 : 736.5999
$\frac{1}{8}$: 3.1750	6 : 152.7800	18 : 457.2000	30 : 761.9999
$\frac{1}{4}$: 6.3500	7 : 177.8000	19 : 482.6000	31 : 787.3999
$\frac{1}{2}$: 12.7000	8 : 203.2000	20 : 508.0000	32 : 812.7999
$\frac{3}{4}$: 19.0500	9 : 228.6000	21 : 533.4000	33 : 838.1999
1 : 25.4000	10 : 254.0000	22 : 558.8000	34 : 863.5999
$1\frac{1}{4}$: 31.7500	11 : 279.4000	23 : 584.1990	35 : 888.9999
$1\frac{1}{2}$: 38.1000	12 : 304.8000	24 : 609.5990	36 : 914.3999
$1\frac{3}{4}$: 44.4500	13 : 330.2000	25 : 634.9999	37 : 939.7999
2 : 50.8000	14 : 355.6000	26 : 660.3999	38 : 965.1999
			39 : 990.5999

2. Feet to metres linear conversion.

Multiply by 0.305. (*Note*. 0.300 is more practical than 0.305 for general calculations.)

ft/m	*ft/m*	*ft/m*
1 : 0.3048	10 : 3.0480	19 : 5.7912
2 : 0.6096	11 : 3.3528	20 : 6.0960
3 : 0.9144	12 : 3.6576	30 : 9.1440
4 : 1.2192	13 : 3.9624	40 : 12.1920
5 : 1.5240	14 : 4.2672	50 : 15.2400
6 : 1.8288	15 : 4.5720	60 : 18.2880
7 : 2.1336	16 : 4.8768	70 : 21.3360
8 : 2.4384	17 : 5.1816	80 : 24.3840
9 : 2.7432	18 : 5.4864	90 : 27.4320
		100 : 30.4800

3. Square feet to square metres.

Multiply by 0.093

sq.ft/sq.m	*sq.ft/sq.m*	*sq.ft/sq.m*	*sq.ft/sq.m*
1 : 0.093	6 : 0.557	20 : 1.858	70 : 6.603
2 : 0.186	7 : 0.650	30 : 2.787	80 : 7.532
3 : 0.279	8 : 0.743	40 : 3.716	90 : 8.461
4 : 0.372	9 : 0.836	50 : 4.645	100 : 9.390
5 : 0.465	10 : 0.929	60 : 5.674	

4. Square yards to square metres.

Multiply by 0.836

sq.yds/sq.m	*sq.yds/sq.m*	*sq.yds/sq.m*	*sq.yds/sq.m*
1 : 0.836	6 : 5.017	20 : 16.722	70 : 58.527
2 : 1.672	7 : 5.853	30 : 25.083	80 : 66.888
3 : 2.508	8 : 6.689	40 : 33.444	90 : 75.249
4 : 3.345	9 : 7.525	50 : 41.805	100 : 83.610
5 : 4.181	10 : 8.361	60 : 50.166	

5. Cubic inches to cubic centimetres.

Multiply by 16.389

cu.in./cu.cm	*cu.in./cu.cm*	*cu.in./cu.cm*	*cu.in./cu.cm*
1 : 16.389	6 : 98.333	20 : 327.778	70 : 1147.223
2 : 32.778	7 : 114.722	30 : 491.667	80 : 1311.112
3 : 49.667	8 : 131.111	40 : 655.556	90 : 1475.001
4 : 65.556	9 : 147.500	50 : 819.445	100 : 1638.890
5 : 81.944	10 : 163.889	60 : 983.334	

6. Cubic feet to cubic metres.

Multiply by 0.028

cu.ft/cu.m.	*cu.ft/cu.m.*	*cu.ft/cu.m.*	*cu.ft/cu.m.*
1 : 0.028	6 : 0.170	20 : 0.566	70 : 1.981
2 : 0.057	7 : 0.198	30 : 0.849	80 : 2.264
3 : 0.085	8 : 0.227	40 : 1.132	90 : 2.547
4 : 0.113	9 : 0.255	50 : 1.415	100 : 2.830
5 : 0.142	10 : 0.283	60 : 1.698	

7. Cubic yards to cubic metres.

Multiply by 0.765

cu.yds/cu.m	*cu.yds/cu.m*	*cu.yds/cu.m*	*cu.yds/cu.m*
1 : 0.765	6 : 4.587	20 : 15.292	70 : 53.522
2 : 1.529	7 : 5.352	30 : 22.938	80 : 61.168
3 : 2.294	8 : 6.116	40 : 30.584	90 : 68.814
4 : 3.058	9 : 6.881	50 : 38.230	100 : 76.460
5 : 3.823	10 : 7.646	60 : 45.876	

8. Pounds to kilograms.

Multiply by 0.454

lb/kg	*lb/kg*	*lb/kg*	*lb/kg*
1 : 0.454	6 : 2.722	20 : 9.072	70 : 31.752
2 : 0.907	7 : 3.175	30 : 13.608	80 : 36.288
3 : 1.361	8 : 3.629	40 : 18.144	90 : 40.824
4 : 1.814	9 : 4.082	50 : 22.680	100 : 45.360
5 : 2.268	10 : 4.536	60 : 27.216	

9. Hundredweight to kilograms.

Multiply by 50.802

cwt/kg	*cwt/kg*	*cwt/kg*	*cwt/kg*
1 : 50.802	6 : 304.814	20 : 1016.046	70 : 3356.171
2 : 101.605	7 : 355.616	30 : 1524.079	80 : 4064.194
3 : 152.407	8 : 406.419	40 : 2032.102	90 : 4572.217
4 : 203.209	9 : 457.221	50 : 2540.125	100 : 5080.240
5 : 254.012	10 : 508.023	60 : 3048.148	

10. Imperial tons to kilograms.

Multiply by 1016.047

tons/kg	*tons/kg*	*tons/kg*	*tons/kg*
1 : 1016.047	6 : 6096.282	20 : 20320.940	70 : 71124.290
2 : 2032.094	7 : 7112.329	30 : 30481.410	80 : 81284.760
3 : 3048.141	8 : 8128.376	40 : 40641.880	90 : 91444.230
4 : 4064.188	9 : 9144.423	50 : 50802.350	100 : 101605.700
5 : 5080.235	10 : 10160.470	60 : 60963.820	

11. Pounds per cubic foot to kilograms per cubic metre.

Multiply by 16.02

$lb/ft^3/kg/m^3$	$lb/ft^3/kg/m^3$	$lb/ft^3/kg/m^3$	$lb/ft^3/kg/m^3$
1 : 16.020	6 : 96.120	20 : 320.400	70 : 1121.400
2 : 32.040	7 : 112.140	30 : 480.600	80 : 1281.600
3 : 48.060	8 : 128.160	40 : 640.800	90 : 1441.800
4 : 64.080	9 : 144.180	50 : 801.000	100 : 1602.000
5 : 80.100	10 : 160.200	60 : 961.200	

12. Pints to litres.

Multiply by 0.568

pt/l	*pt/l*	*pt/l*	*pt/l*
1 : 0.568	6 : 3.408	20 : 11.360	70 : 39.760
2 : 1.136	7 : 3.976	30 : 17.040	80 : 45.440
3 : 1.704	8 : 4.544	40 : 22.720	90 : 51.120
4 : 2.272	9 : 5.112	50 : 28.400	100 : 56.800
5 : 2.840	10 : 5.680	60 : 34.080	

13. Gallons to litres.

Multiply by 4.544

gall/l	*gall/l*	*gall/l*	*gall/l*
1 : 4.544	6 : 27.262	20 : 90.872	70 : 318.052
2 : 9.087	7 : 31.805	30 : 136.308	80 : 363.488
3 : 13.631	8 : 36.349	40 : 181.744	90 : 408.924
4 : 18.174	9 : 40.892	50 : 227.180	100 : 454.400
5 : 22.718	10 : 45.436	60 : 272.616	

14. Miles per hour to kilometres per hour.

Multiply by 0.6214

m.p.h./k.p.h.	*m.p.h./k.p.h.*	*m.p.h./k.p.h.*	*m.p.h./k.p.h.*
1 : 0.6124	6 : 3.7284	20 : 12.4270	70 : 43.4960
2 : 1.2428	7 : 4.3498	30 : 18.6410	80 : 49.7100
3 : 1.8642	8 : 4.9712	40 : 24.8550	90 : 55.9230
4 : 2.4856	9 : 5.5926	50 : 31.0690	100 : 62.1370
5 : 3.1070	10 : 6.2140	60 : 37.2820	

20 Answers

Addition

1. 10 2. 20 3. 11 4. 18 5. 30 6. 10.5
7. 10 8. 30 9. 13 10. 12.5.

Subtraction

1. 3 2. 7 3. 30 4. 20 5. 14 6. 5.5
7. 10.25 8. 60 9. 2 10. 10.04.

Division

1. 2 2. 5 3. 9 4. 3 5. 3 6. 4
7. 6.5 8. 9.524 9. 4.25 10. 3.018.

Fractions to decimals

1. 0.0265 2. 0.0313 3. 0.125 4. 0.6 5. 0.1875
6. 0.172 7. 0.388 8. 0.094 9. 0.3125 10. 0.25

Ratios

1. 1:16 2. 1:4 3. 3:4 4. 7:8 5. 11:16 6. 23.4
7. 50:7 8. 23:6 9. 8:48 or 1:6 10. 15.35

Percentages

1. 1% 2. 10% 3. 12.5% 4. 20% 5. 150%
6. £13.20 7. £5.86½ 8. £8.10 9. £15.45
10. £42.48.

Averages

1. 16.2 2. 9.617 3. 1.771.

Rearranging formulae

1. $A = BC$ 2. $B = A + C$ 3. $C = \dfrac{AB}{D}$ 4. $D = \dfrac{AB}{C}$

5. $6 = 8 - 2$ 6. $A^2 4 = D$ 7. $C = \dfrac{\sqrt{A}}{D}$ 8. $A = \dfrac{4}{6^2}$

9. $H = \dfrac{q^2 L}{(3D^5)}$ 10. $L = \dfrac{3D^5 H}{q^2}$.

Pythagoras' theorem

1. 7.81 m 2. 5.196 m 3. 6.245 m

Square roots using tables

1. 2.236 2. 4.123 3. 2.510 4. 4.519 5. 25.00
6. 20.00 7. 7.000 8. 0.7746 9. 9.685 10. 10.00

Moments

1. 2.5 kgf and 24.525 newtons.
1a. 10 kgf and 98.1 newtons.
2. 160 kgf and 1569.6 newtons.

Levers

1. 155.56 kgf and 1525.99 newtons.
2. 666.67 kgf and 6539.99 newtons.

Water pressure

1. 12,000 kgf and 117,720 newtons.
2. 294.5 kgf and 2889 newtons.

Box's formula

1. 1.2 litres/second 2. 6.87 m.

Chezy formula

1. 1.84 litres/second. 2. 1.9.

Logarithms

1. 1.0 2. $\bar{1}.0$ 3. 3 4. 0.2553 5. 0.6021
6. 9085 7. 0.6551 8. 0.1393 9. 1585
10. 2423 11. 1091 12. 396.7 13. 4.2491
14. 11.77 15. 9.359 16. 3.783 17. 9.539